TOPICS IN REGRESSION ANALYSIS

Macmillan Series in Economics
Lawrence R. Klein, Consulting Editor

WITHDRAWI

TOPICS
IN REGRESSION
ANALYSIS

ARTHUR S. GOLDBERGER

Professor of Economics, University of Wisconsin

The Macmillan Company
NEW YORK

Collier-Macmillan Limited
LONDON

© Copyright, Arthur S. Goldberger, 1968

First Printing

Library of Congress catalog card number: 68–15265

THE MACMILLAN COMPANY, NEW YORK
COLLIER-MACMILLAN CANADA, LTD., TORONTO, ONTARIO

PRINTED IN THE UNITED STATES OF AMERICA

In memory of Stefan Valavanis

Preface

THIS BOOK IS BASED ON A COURSE OF EIGHT LECTURES given at the Center of Planning and Economic Research, Athens, Greece, in the spring of 1965. The course did not attempt to provide a systematic development of the theory and practice of regression analysis. Rather, the objective was to discuss a selection of topics from the point of view of a researcher who is interested in using this important tool in a productive way in empirical investigations. The presumption was made that the participants in the course had some familiarity with the principles of multiple regression and some experience with its use in practice. Some knowledge of the elements of statistical inference and of matrix algebra was also taken for granted. Similar presumptions are made for readers of this book.

Regression analysis is, of course, a thoroughly documented subject and the present book cannot be considered a totally original contribution. Nevertheless, attention is called to the development of the estimation of conditional expectations in Chapter 2 and to the heuristic approach to measuring precision of regression coefficients in Chapters 5 and 6; if not original, this material is at least unconventional.

The course was given toward the end of my academic year's stay at the Center of Planning and Economic Research as Visiting Professor, University of California (Berkeley-Athens Project), on leave from the University of Wisconsin. My thanks go to these institutions. Less abstractly, I am grateful to Guy H. Orcutt for discussions over many years which enriched my appreciation of the conditional expectation concept; to Donald Bear, Louis Lefeber, Harry Kelejian, and Hans Theil for guidance on specific problems; and to Theodore Gamaletsos and Alan Duchan for comments and corrections. A large measure of thanks goes to Jan Kmenta, whose careful critique of the entire manuscript compelled me to rework many points (although perhaps not as many as he wished). For expert typing of difficult material I am grateful, once again, to Mrs. Alice Wilcox and to Miss Linda Anderson.

    Last but not least, I am happy to acknowledge my indebtness to Pan A. Yotopoulos for encouragement in the preparation of the course and the book.

A. S. G.

Contents

TOPICS IN REGRESSION ANALYSIS

Chapter 1   Introduction

MOST ECONOMIC THEORY IS CONCERNED WITH relationships among variables. The quantity of a commodity demanded by an individual consumer is related to his income and the price of the commodity; national consumption expenditures are related to current and past values of national income; and so forth. Correspondingly, most of empirical economics is concerned with measuring relationships among variables, that is, with the quantification of the abstract relations of economic theory.

The most commonly used tool for measuring economic relationships is linear regression analysis, which is the concern of this book. No systematic development of linear regression analysis will be attempted here, however. Rather we shall explore some selected topics of that rather broad subject. Thus it will be presumed that the reader is already familiar with the theory of linear regression analysis—as developed in, for example, Johnston (1963, Chap. 4), as well with some of its empirical applications— as presented in, for example, Ezekiel and Fox (1959). Some knowledge is also presumed of the elements of probability distribution theory (joint, marginal, and conditional distributions, in particular), mathematical expectations, and sampling theory—as developed in, for example, Mood and Graybill (1963, Chaps. 1–8). Finally, at various points matrix algebra is employed—a convenient reference for this might be Johnston (1963, Chap. 3).

1

The plan of this book is as follows. In Chapter 2 the concept of the population regression function as a key feature of a stochastic relationship is developed and the sample regression function is proposed as its estimator. Chapter 3 considers the interpretation of the regression coefficients obtained when a linear regression function is fit to a body of data. A related interpretation of the coefficient of determination is attempted in Chapter 4. Some bases for judging the reliability of regression coefficients are presented in Chapter 5, and Chapter 6 is devoted to further analysis of this problem and to the related question of multicollinearity. The treatment up to this point proceeds with a minimum of formal statistical theory, a limitation which is removed in Chapter 7. Chapter 8 serves to establish that linear regression analysis is in fact applicable in a wide variety of nonlinear problems. Finally, Chapter 9 offers some suggestions for the selection of appropriate functional forms.

Chapter 2   Relationships and
Regressions

2.1. INTRODUCTION

To START, WE SHOULD DISTINGUISH BETWEEN THE deterministic relationships usually employed in economic theory and the stochastic relationships relevant in empirical economics. We say that there is a *deterministic* relationship between a variable y and the variables $x_1, \ldots, x_K$ if, corresponding to each set of values of the x's, there is a unique value of y, namely, $y = f(x_1, \ldots, x_K)$. Further we say that y is dependent upon the x's if $f(x_1, \ldots, x_K)$ is not constant over all sets of values of the x's. On the other hand, we say that there is a *stochastic* relationship between a variable y and the variables $x_1, \ldots, x_K$ if, corresponding to each set of values of the x's, there is a conditional probability distribution of values of y, namely, $p(y \mid x_1, \ldots, x_K) = f(x_1, \ldots, x_K)$. In this case we say that y is dependent upon the x's if $f(x_1, \ldots, x_K)$ is not constant over all sets of values of the x's.

In principle, specifying a stochastic relationship means specifying the full set of conditional probability distributions. In practice, however, most interest is attached to a few parameters of these distributions. In particular, interest is very often confined to the *population regression function* $E(y \mid x_1, \ldots, x_K) = g(x_1, \ldots, x_K)$, which describes how the *average* value of y varies with the x's.

Regression analysis is essentially concerned with estimation of such a population regression function on the basis of a sample of observations

3

drawn from the joint probability distribution of y, $x_1, \ldots, x_K$. Now, in general, the selection of an appropriate method of estimation requires a rather detailed set of assumptions on the mechanism which generates the sample, along with reliance on some criteria of statistical inference. For present purposes, however, we shall draw on the intuitively plausible *analogy principle of estimation*, which proposes that population parameters be estimated by sample statistics which have the same property in the sample as the parameters do in the population. For example, in the case of a univariate distribution, this principle might propose the sample mean as the estimator of the population mean, the sample variance as the estimator of the population variance, and so forth.

Since $E(y \mid x_1, \ldots, x_K)$ gives the population mean value of y conditional upon $x_1, \ldots, x_K$, the analogy principle would seem to suggest the *cell-mean function* as the estimator of the population regression function. That is, consider a subset of sample observations (= cell) defined by a particular set of values of the x's, and take the mean value of y in this cell as the estimator of the conditional expectation of y for that set of x's. Unfortunately, this approach is often impracticable, because the sample contains insufficient observations in many cells. Indeed, we may be interested in estimating the population regression function over sets of values of the x's which do not appear in our sample at all.

Still, the analogy principle can be modified a bit to provide an approach to estimating population regression functions that is widely practicable. As we shall see, a characteristic property of the population regression function is that the deviations from it are uncorrelated *in the population* with the conditioning variables. Suppose that we know the mathematical form of the population regression function—say, $E(y \mid x_1, \ldots, x_K)$ is a linear function of the x's. Then the modified analogy principle suggests that we estimate the population regression function by a sample regression function of the same mathematical form, chosen to make the deviations from it uncorrelated *in the sample* with the conditioning variables.

2.2. RELATIONSHIPS AND REGRESSIONS: ONE CONDITIONING VARIABLE

To clarify the position, let us first consider the case of a single conditioning variable. A stochastic relationship between the dependent variable y and the conditioning variable x is fully specified by a tabulation of the conditional probability distributions $p(y \mid x)$ for all possible values of x, as in the following tabulation:

			x		
y	x_1	$\cdots$	x_j	$\cdots$	x_J
y_1	p_{11}	$\cdots$	p_{1j}	$\cdots$	p_{1J}
$\vdots$	$\vdots$		$\vdots$		$\vdots$
y_i	p_{i1}	$\cdots$	p_{ij}	$\cdots$	p_{iJ}
$\vdots$	$\vdots$		$\vdots$		$\vdots$
y_I	p_{I1}	$\cdots$	p_{Ij}	$\cdots$	p_{IJ}

Here $p_{ij} = p(y_i \mid x_j)$ denotes the probability that y takes on the value y_i given that ($=$ conditional upon the event that) x takes on the value x_j. For simplicity we treat the case of discrete distributions, in which there are I distinct values of y and J distinct values of x. The extension to the case of continuous distributions is straightforward. Note that each column provides a probability distribution; each column sum is unity: $\sum_{i=1}^{I} p_{ij} = 1$ ($j = 1, \ldots, J$).

The conditional expectation of y given x_j is the mean of the jth distribution: $E(y \mid x_j) = \sum_{i=1}^{I} y_i p_{ij}$. The general expression for the population regression function may then be written $E(y \mid x) = \sum_y y p(y \mid x)$, where $\sum_y$ denotes summation over all possible values of y. Similarly, the conditional variance of y given x_j is the variance of the jth distribution: $V(y \mid x_j) = E\{[y - E(y \mid x_j)]^2 \mid x_j\} = \sum_{i=1}^{I} [y_i - E(y \mid x_j)]^2 p_{ij}$, and the general expression for the so-called population skedastic function may be written $V(y \mid x) = E\{[y - E(y \mid x)]^2 \mid x\} = \sum_y [y - E(y \mid x)]^2 p(y \mid x)$. Other conditional parameters are similarly defined.

We shall say that y is *stochastically independent* of x if and only if the conditional probability distribution of y is the same for all values of x, that is, if and only if $p(y \mid x)$ is constant over x, that is, if and only if all the columns of the above tabulation are identical. In that case, clearly, the conditional expectation is the same for all values of x, the conditional variance is the same for all values of x, and similarly for all other conditional parameters. Note that this formal property of stochastic independence conforms to the usage of Section 2.1: When y is stochastically independent of x, then y is not "dependent upon x."

We proceed to state some elementary properties of conditional expectations which are required in the sequel. They are rather immediate consequences of the fact that an expectation is simply a (population) average; for a more formal discussion, see, for example, Mood and Graybill (1963,

Chap. 5) or Goldberger (1964, pp. 79–87). First, we have the simple effect of an additive or multiplicative constant upon the conditional expectation of a random variable:

If c is a constant given x, then $E[(y + c)|x] = E(y|x) + c$
and $E(cy|x) = cE(y|x)$.   (2.1)

Next the unconditional expectation of y is defined as $E(y) = \sum_{i=1}^{I} y_i p(y_i)$, where $p(y_i)$ is the unconditional (=marginal) probability that y takes on the value y_i. Clearly, $p(y_i) = \sum_{j=1}^{J} p_{ij} p(x_j)$, where $p(x_j)$ is the unconditional probability that x takes on the value x_j. It follows that the unconditional expectation of y is simply the expectation of its conditional expectations:

$$E(y) = E[E(y|x)] = \sum_{j=1}^{J} E(y|x_j)p(x_j). \qquad (2.2)$$

Suppose that the conditional expectation of y is the same for all values of x. Then the unconditional expectation of y has that same common value:

If $E(y|x)$ is constant over all x, then $E(y) = E(y|x)$.   (2.3)

Here the population regression function is constant; we may say that y is *mean-independent* of x. Now, two variables are said to be uncorrelated if and only if their covariance is zero, that is, if and only if the expectation of their product is the product of their expectations. Clearly, if y is mean-independent of x, then y and x are *uncorrelated*:

If $E(y|x)$ is constant over all x, then $E(yx) = E(y)E(x)$.   (2.4)

Proof: The unconditional expectation of yx is obtainable as the expectation of its conditional expectations:

$$E(yx) = \sum_{j=1}^{J} E(yx_j|x_j)p(x_j).$$

Given x_j, x_j is constant, so that

$$E(yx_j|x_j) = x_j E(y|x_j).$$

Since $E(y|x)$ is constant over x,

$$E(y|x_j) = E(y).$$

Therefore,

$$E(yx) = \sum_{j=1}^{J} x_j E(y) p(x_j) = E(y) \sum_{j=1}^{J} x_j p(x_j)$$

$$= E(y)E(x).$$

It is instructive to note that while stochastic independence implies mean independence, the converse is not true, and while mean independence implies uncorrelatedness, the converse is not true.

Armed with these general results, we can see how the population regression function is characterized by certain properties of the deviations from it. We denote these deviations by ε and refer to them as *disturbances*. The disturbances have zero expectation conditional on every value of the conditioning variable:

$$\text{If } \varepsilon = y - E(y \mid x), \text{ then } E(\varepsilon \mid x) = 0 \text{ for all } x. \tag{2.5}$$

Proof: Since $E(y \mid x)$ is a constant given x,

$$E(\varepsilon \mid x) = E\{[y - E(y \mid x)] \mid x\} = E(y \mid x) - E(y \mid x) = 0.$$

This feature—that the deviations from the population regression function have zero expectation conditional upon every value of the conditioning variable—is indeed distinctive. For let $g^*(x)$ be an arbitrary function of the conditioning variable and let $\varepsilon^* = y - g^*(x)$ be the deviations from it. Then $E(\varepsilon^* \mid x) = E(y \mid x) - g^*(x)$ will not be zero for all values of x, unless $g^*(x) \equiv E(y \mid x)$.

The cell-mean function has the analogous feature in the sample: In each cell the deviations from the cell mean sum to zero and hence average out to zero. But as suggested earlier, the cell-mean approach is generally impracticable, so that we should look for some weaker properties of the population regression function. If in (2.3) and (2.4) we let ε take the role of y, then we can deduce, from (2.5),

$$\text{If } \varepsilon = y - E(y \mid x), \text{ then } E(\varepsilon) = 0 \text{ and } E(\varepsilon x) = 0. \tag{2.6}$$

Thus the deviations from the population regression function have zero unconditional expectation and are uncorrelated with the conditioning variable. These are indeed weaker properties: An unconditional expectation can be zero without each of the conditional expectations being zero, and constancy of the conditional expectation, while sufficient, is not necessary for uncorrelatedness. There are functions other than the population regression function which have these weaker properties in the population. Correspondingly, in a sample, not only the cell-mean function, but other functions as well, will have the analogous properties: deviations summing to zero overall (not necessarily within any cell) and being uncorrelated with the conditioning variable (not necessarily mean-independent of it). If we know or are willing to assume the mathematical form of the population regression function, then the modified analogy principle suggests that we estimate it by a function of the same mathematical form which has the weaker analogous properties. This proposal is in general operational.

To illustrate this point, suppose that we know or are willing to assume that the population regression function is linear:

$$E(y \mid x) = \beta_0 + \beta_1 x. \tag{2.7}$$

Writing $y = \beta_0 + \beta_1 x + \varepsilon$, we have, from (2.6), these properties for the disturbance ε:

$$E(\varepsilon) = 0, \qquad E(\varepsilon x) = 0. \tag{2.8}$$

Given a sample of T joint observations y_t, x_t ($t = 1, \ldots, T$), the modified analogy principle now leads us to seek a sample function which has the same mathematical form as (2.7) and the deviations from which have in the sample the properties analogous to (2.8). Thus we write

$$y_t = b_0 + b_1 x_t + e_t \qquad (t = 1, \ldots, T), \tag{2.9}$$

in which $b_0 + b_1 x$ denotes the function we seek and e a deviation from it. These deviations are to have the properties

$$\sum_{t=1}^{T} e_t = 0, \qquad \sum_{t=1}^{T} e_t x_t = 0, \tag{2.10}$$

which in view of (2.9) are equivalent to

$$\sum_{t=1}^{T} (y_t - b_0 - b_1 x_t) = 0,$$

$$\sum_{t=1}^{T} [(y_t - b_0 - b_1 x_t) x_t] = 0. \tag{2.11}$$

This last pair of equations determines our linear sample regression function. In a more compact and familiar form we write the pair

$$T b_0 + \sum x \, b_1 = \sum y,$$
$$\sum x \, b_0 + \sum x^2 \, b_1 = \sum xy, \qquad (2.12)$$

where here and in what follows $\sum$ is shorthand for $\sum_{t=1}^{T}$ (unless otherwise indicated) and the observation subscript t has been suppressed.

The solution of (2.12) is readily obtained; the slope of the linear sample regression function is

$$b_1 = \frac{T \sum xy - \sum x \sum y}{T \sum x^2 - (\sum x)^2} = \frac{\sum (x - \bar{x})(y - \bar{y})}{\sum (x - \bar{x})^2}, \qquad (2.13)$$

and the intercept of the linear sample regression function is

$$b_0 = (\sum y - b_1 \sum x)/T = \bar{y} - b_1 \bar{x}, \qquad (2.14)$$

where here and in what follows the bar superscript denotes the sample mean: $\bar{x} = \sum_{t=1}^{T} x_t/T$ and $\bar{y} = \sum_{t=1}^{T} y_t/T$.

The values of the sample regression function corresponding to sample values of x are called the *calculated values* of y, are denoted by $\hat{y}$, and are computed as

$$\hat{y}_t = b_0 + b_1 x_t \qquad (t = 1, \ldots, T). \qquad (2.15)$$

The deviations of y from $\hat{y}$ in the sample are called the *residuals*, are denoted by e, and are computed as

$$e_t = y_t - \hat{y}_t \qquad (t = 1, \ldots, T). \qquad (2.16)$$

It is readily confirmed that

$$\sum e = \sum xe = 0, \qquad (2.17)$$

whence

$$\sum \hat{y} = \sum (y - e) = \sum y, \qquad \sum \hat{y}e = \sum (b_0 + b_1 x)e = 0, \qquad (2.18)$$

$$\sum y^2 = \sum (\hat{y} + e)^2 = \sum \hat{y}^2 + \sum e^2. \qquad (2.19)$$

Example 2.1

Consider the stochastic relationship defined by the following tabulation of $p(y\,|\,x)$.

	x	
y	-1	1
-1	0.2	0.1
0	0.1	0.3
1	0.7	0.6
	1.0	1.0

The conditional means and variances are

$E(y\,|\,x=-1)=(-1)(0.2)+(0)(0.1)+(1)(0.7)=0.5,$

$E(y\,|\,x=1)=(-1)(0.1)+(0)(0.3)+(1)(0.6)=0.5,$

$V(y\,|\,x=-1)=(-1-0.5)^2(0.2)+(0-0.5)^2(0.1)+(1-0.5)^2(0.7)=0.65,$

$V(y\,|\,x=1)=(-1-0.5)^2(0.1)+(0-0.5)^2(0.3)+(1-0.5)^2(0.6)=0.45.$

Note that y is not stochastically independent of x (the two columns are not identical) but is mean-independent of x: $E(y\,|\,x)=0.5$ for all possible values of x.

Example 2.2

Consider the stochastic relationship defined by the following tabulation of $p(y\,|\,x)$, given by rows rather than columns for compactness.

$x=-1:\ p(y=-1\,|\,x=-1)=\tfrac{1}{2}=p(y=3\,|\,x=-1),$

$x=0:\ p(y=0\,|\,x=0)=\tfrac{1}{2}=p(y=4\,|\,x=0),$

$x=1:\ p(y=0\,|\,x=1)=\tfrac{1}{2}=p(y=2\,|\,x=1).$

The conditional means are:

$E(y\,|\,x=-1)=(-1)(\tfrac{1}{2})+(3)(\tfrac{1}{2})=1,$

$E(y\,|\,x=0)=(0)(\tfrac{1}{2})+(4)(\tfrac{1}{2})=2,$

$E(y\,|\,x=1)=(0)(\tfrac{1}{2})+(2)(\tfrac{1}{2})=1.$

Further we have

$$E(yx \mid x = -1) = (-1)E(y \mid x = -1) = -1,$$
$$E(yx \mid x = 0) = (0)E(y \mid x = 0) = 0,$$
$$E(yx \mid x = 1) = (1)E(y \mid x = 1) = 1.$$

Now suppose that the marginal distribution of x is $p(x = -1) = p(x = 0) = p(x = 1) = \frac{1}{3}$; this implies, for the unconditional expectations,

$$E(x) = (-1)p(x = -1) + (0)p(x = 0) + (1)p(x = 1)$$
$$= (-1)(\tfrac{1}{3}) + (0)(\tfrac{1}{3}) + (1)(\tfrac{1}{3}) = 0,$$

$$E(y) = E(y \mid x = -1)p(x = -1) + E(y \mid x = 0)p(x = 0)$$
$$+ E(y \mid x = 1)p(x = 1)$$
$$= (1)(\tfrac{1}{3}) + (2)(\tfrac{1}{3}) + (1)(\tfrac{1}{3}) = \tfrac{4}{3},$$

$$E(yx) = E(yx \mid x = -1)p(x = -1) + E(yx \mid x = 0)p(x = 0)$$
$$+ E(yx \mid x = 1)p(x = 1)$$
$$= (-1)(\tfrac{1}{3}) + (0)(\tfrac{1}{3}) + (1)(\tfrac{1}{3}) = 0.$$

Note that y is not stochastically independent of x, or mean-independent of x, but y and x are uncorrelated: $E(yx) = E(y)E(x)$.

Example 2.3

Continuing to use the data of the previous example, let ε denote the deviations from the population regression function. The following tabulation of $p(\varepsilon \mid x)$ follows from that of $p(y \mid x)$:

$$x = -1: \ p(\varepsilon = -2 \mid x = -1) = \tfrac{1}{2} = p(\varepsilon = 2 \mid x = -1),$$
$$x = 0: \ p(\varepsilon = -2 \mid x = 0) = \tfrac{1}{2} = p(\varepsilon = 2 \mid x = 0),$$
$$x = 1: \ p(\varepsilon = -1 \mid x = 1) = \tfrac{1}{2} = p(\varepsilon = 1 \mid x = 1).$$

Note that $E(\varepsilon \mid x) = 0$ for all three possible values of x, that $E(\varepsilon) = 0$, and that $E(\varepsilon x) = 0$. But the population regression function is not the only one whose deviations have the latter two properties. For example, take the function

$$g^*(x) = \begin{cases} 0 & \text{if } x = -1 \\ 4 & \text{if } x = 0 \\ 0 & \text{if } x = 1 \end{cases}$$

and let $\varepsilon^* = y - g^*(x)$ denote the deviations from it. The following tabulation of $p(\varepsilon^* \mid x)$ follows from that of $p(y \mid x)$:

$$x = -1: \ p(\varepsilon^* = -1 \mid x = -1) = \tfrac{1}{2} = p(\varepsilon^* = 3 \mid x = -1),$$

$$x = \ \ \ 0: \ p(\varepsilon^* = -4 \mid x = \ \ \ 0) = \tfrac{1}{2} = p(\varepsilon^* = 0 \mid x = \ \ \ 0),$$

$$x = \ \ \ 1: \ p(\varepsilon^* = \ \ \ 0 \mid x = \ \ \ 1) = \tfrac{1}{2} = p(\varepsilon^* = 2 \mid x = \ \ \ 1).$$

The conditional expectations of ε^* are thus

$$E(\varepsilon^* \mid x = -1) = 1, \qquad E(\varepsilon^* \mid x = 0) = -2, \qquad E(\varepsilon^* \mid x = 1) = 1,$$

while those of $\varepsilon^* x$ are

$$E(\varepsilon^* x \mid x = -1) = -1, \qquad E(\varepsilon^* x \mid x = 0) = 0, \qquad E(\varepsilon^* x \mid x = 1) = 1.$$

Then with $p(x = -1) = p(x = 0) = p(x = 1) = \tfrac{1}{3}$, we have

$$E(\varepsilon^*) \ = (1)(\tfrac{1}{3}) + (-2)(\tfrac{1}{3}) + (1)(\tfrac{1}{3}) = 0,$$

$$E(\varepsilon^* x) = (-1)(\tfrac{1}{3}) + (0)(\tfrac{1}{3}) + (1)(\tfrac{1}{3}) = 0,$$

although $E(\varepsilon^* \mid x) \neq 0$ for any value of x.

Example 2.4

For the set of observations y_t, x_t ($t = 1, \ldots, 6$) given below, the calculation of the linear sample regression function is shown along with numerical checks of its properties. A scatter diagram of these observations is given in Figure 2.1.

t	x	y
1	−1	0
2	0	2
3	1	1
4	2	2
5	0	−1
6	0	1

$$T = 6 \qquad \sum x = 2 \qquad \sum y = 5$$
$$\sum x^2 = 6 \qquad \sum xy = 5$$
$$\sum y^2 = 11$$

$$b_1 = \frac{(6)(5) - (2)(5)}{(6)(6) - (2)^2} = \frac{5}{8}$$

$$b_0 = [5 - (5/8)(2)]/6 = \tfrac{5}{8}$$

x	y	$\hat{y}$	e	xe	$\hat{y}e$	y^2	$\hat{y}^2$	e^2
-1	0	0/8	0/8	$-0/8$	0/64	0	0/64	0/64
0	2	5/8	11/8	0/8	55/64	4	25/64	121/64
1	1	10/8	$-2/8$	$-2/8$	$-20/64$	1	100/64	4/64
2	2	15/8	1/8	2/8	15/64	4	225/64	1/64
0	-1	5/8	$-13/8$	0/8	$-65/64$	1	25/64	169/64
0	1	5/8	3/8	0/8	15/64	1	25/64	9/64
Sums 2	5	40/8 $=5$	0/8 $=0$	0/8 $=0$	0/64 $=0$	11 $=704/64$	400/64	304/64

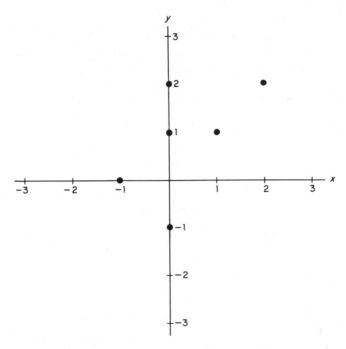

Figure 2.1 Scatter diagram of y on x.

2.3. RELATIONSHIPS AND REGRESSIONS: *K* CONDITIONING VARIABLES

It is a straightforward matter to generalize the discussion of Section 2.2 to the case of a stochastic relationship between a variable y and a set of conditioning variables $x_1, \ldots, x_K$. Again confining our attention to the discrete case, the relationship is fully specified by a tabulation of the conditional probability distributions $p(y \mid x_1, \ldots, x_K)$, one column for each set of values of the x's. (*Caution*: The subscripts on the x's now denote different variables rather than different values of the same variable.) Letting asterisks denote particular values, we have $p(y_i \mid x_{1*}, \ldots, x_{K*})$ for the probability that y takes on the value y_i given that (=conditional upon the event that) the x's take on the values $x_1 = x_{1*}, \ldots, x_K = x_{K*}$. The conditional expectation of y given $x_{1*}, \ldots, x_{K*}$ is the column mean, $E(y \mid x_{1*}, \ldots, x_{K*}) = \sum_{i=1}^{I} y_i p(y_i \mid x_{1*}, \ldots, x_{K*})$. The general expression for the population regression function may be written $E(y \mid x_1, \ldots, x_K) = \sum_y y p(y \mid x_1, \ldots, x_K)$. The conditional variance and other conditional parameters are similarly defined.

A more compact notation will be helpful. Let $x' = (x_1, \ldots, x_K)$ denote the set of K conditioning variables and $x'_* = (x_{1*}, \ldots, x_{K*})$ a particular set of values of them. Then we may write $p(y \mid x')$ for the conditional probability distribution, $p(y_i \mid x'_*)$ for the probability that $y = y_i$ given that $x' = x'_*$, $E(y \mid x'_*) = \sum_{i=1}^{I} y_i p(y_i \mid x'_*)$ for the conditional expectation of y given x'_*, and $E(y \mid x') = \sum_y y p(y \mid x')$ for the population regression function. We shall have occasion to consider arbitrary partitionings of x' into two mutually exclusive and exhaustive subsets of variables, to be denoted x'_A and x'_B, and to refer to a single arbitrary element of x' as x_k.

We shall say that y is stochastically independent of x'—that is, of the set of variables $x_1, \ldots, x_K$—if and only if the conditional probability distribution of y is the same for all values of x', that is, if and only if $p(y \mid x')$ is constant over x', that is, if and only if all the columns are identical. In that case, clearly, the conditional expectation is the same for all values of x' and so are the conditional variance and the other conditional parameters. Note that if y is independent of a set of variables, then it is independent of any subset of them:

> If y is stochastically independent of x',
> then y is stochastically independent of x'_A. (2.20)

Proof: First, $p(y \mid x'_A) = \sum_{x'_B} p(y \mid x') p(x'_B)$, where $p(x'_B)$ denotes the probability of, and $\sum_{x'_B}$ denotes summation over all possible values of, the complementary subset of variables x'_B. Then if $p(y \mid x')$ is constant [and hence equal to $p(y)$], we have $p(y \mid x'_A) = p(y) \sum_{x'_B} p(x'_B) = p(y)$, so that $p(y \mid x'_A)$ is indeed constant over x'_A.

Also, for future reference, we note the definition of independence between two sets of variables: We say that the set of variables $y_1, \ldots, y_M$ is independent of the set of variables $x_1, \ldots, x_K$ if and only if the (joint) conditional probability distribution $p(y_1, \ldots, y_M \mid x_1, \ldots, x_K)$ is the same for all values of $x_1, \ldots, x_K$.

We may now extend the elementary results of Section 2.2. The effect of an additive or multiplicative constant upon a conditional expectation is given by the following:

If c is a constant given x', then $E[(y + c)|x'] = E(y|x') + c$
and $E(cy|x') = cE(y|x')$.   (2.21)

Next the conditional expectation of y given x'_A is defined by $E(y|x'_A) = \sum_{i=1}^{I} y_i p(y_i|x'_A)$; this "partial" conditional expectation is simply the expectation of conditional expectations:

$$E(y|x'_A) = E_{x'_B}[E(y|x')] = \sum_{x'_B} [E(y|x')]p(x'_B),   (2.22)$$

where, as the last expression indicates, $E_{x'_B}$ denotes the expectation over only x'_B. Similarly, the unconditional expectation of y is an expectation of conditional expectations:

$$E(y) = E[E(y|x')] = \sum_{x'} [E(y|x')]p(x').   (2.23)$$

Suppose that the conditional expectation is the same for all values of x'. Then the unconditional expectation has that same common value:

If $E(y|x')$ is constant over all x', then $E(y) = E(y|x')$;   (2.24)

and, similarly:

If $E(y|x')$ is constant over all x', then $E(y|x'_A) = E(y)$,
in particular, $E(y|x_k) = E(y)$.   (2.25)

Again, constancy of the conditional expectation implies uncorrelatedness:

If $E(y|x')$ is constant over all x', then $E(yx_k) = E(y)E(x_k)$
for any k.   (2.26)

Once again it should be recognized that stochastic independence of y and x' is sufficient but not necessary for $E(y|x')$ to be constant, and that constancy of $E(y|x_k)$ is sufficient but not necessary for $E(yx_k) = E(y)E(x_k)$.

As before, these general results imply some characteristic features for a population regression function. Denoting the deviations from the population regression function, the disturbances, by $\varepsilon = y - E(y|x')$, we see that they have zero expectation conditional upon every set of values of the conditioning variables:

$$\text{If } \varepsilon = y - E(y \mid x'), \text{ then } E(\varepsilon \mid x') = 0 \text{ for all } x'. \tag{2.27}$$

The population regression function is the only function whose deviations have this property in the population, and the cell-mean function is the only one which has the analogous property in the sample. The weaker properties of the population regression function are as follows:

$$\text{If } \varepsilon = y - E(y \mid x'), \text{ then } E(\varepsilon) = 0 \text{ and } E(\varepsilon x_k) = 0 \text{ for all } k; \tag{2.28}$$

these follow from (2.27) if we let ε take the role of y in (2.24) and (2.26). Not only the cell-mean function, but other functions as well, will have the weaker analogous properties in the sample. If we know or are willing to assume the mathematical form of the population regression function, then the modified analogy principle suggests that we estimate it by a function of the same mathematical form which has the weaker analogous properties, a proposal which is in general operational.

To illustrate, suppose that we know or are willing to assume that the population regression function is linear:

$$E(y \mid x_1, \ldots, x_K) = \beta_0 + \beta_1 x_1 + \cdots + \beta_K x_K. \tag{2.29}$$

Writing $y = \beta_0 + \beta_1 x_1 + \cdots + \beta_K x_K + \varepsilon$, we have, from (2.28), these properties for the disturbance ε:

$$E(\varepsilon) = 0, \ E(\varepsilon x_1) = 0, \ldots, E(\varepsilon x_K) = 0. \tag{2.30}$$

Given a sample of T joint observations $y_t, x_{t1}, \ldots, x_{tK}$ ($t = 1, \ldots, T$), the modified analogy principle now leads us to seek a sample function which has the same mathematical form as (2.29) and the deviations from which have in the sample the properties analogous to (2.30). Thus we write

$$y_t = b_0 + b_1 x_{t1} + \cdots + b_K x_{tK} + e_t \quad (t = 1, \ldots, T), \tag{2.31}$$

in which $b_0 + b_1 x_1 + \cdots + b_K x_K$ denotes the function we seek and e the deviations from it. These deviations are to have the properties

$$\sum_{t=1}^{T} e_t = 0, \ \sum_{t=1}^{T} e_t x_{t1} = 0, \ldots, \sum_{t=1}^{T} e_t x_{tK} = 0, \tag{2.32}$$

which in view of (2.31) are equivalent to

$$\sum_{t=1}^{T} (y_t - b_0 - b_1 x_{t1} - \cdots - b_K x_{tK}) = 0$$

$$\sum_{t=1}^{T} [(y_t - b_0 - b_1 x_{t1} - \cdots - b_K x_{tK})x_{t1}] = 0$$

$$\cdots \qquad \cdots \qquad (2.33)$$

$$\sum_{t=1}^{T} [(y_t - b_0 - b_1 x_{t1} - \cdots - b_K x_{tK})x_{tK}] = 0.$$

This system of equations determines our linear sample regression function. In a more compact and familiar form we write the system as

$$T \quad b_0 + \sum x_1 \quad b_1 + \cdots + \sum x_K \quad b_K = \sum y$$

$$\sum x_1 b_0 + \sum x_1^2 \quad b_1 + \cdots + \sum x_1 x_K b_K = \sum x_1 y$$

$$\cdots \qquad \cdots \qquad \cdots \qquad \cdots \qquad (2.34)$$

$$\sum x_K b_0 + \sum x_K x_1 b_1 + \cdots + \sum x_K^2 \quad b_K = \sum x_K y.$$

The solution to this set of $1 + K$ simultaneous linear equations will give the intercept b_0 and slopes $b_1, \ldots, b_K$ of the sample regression function. Although it is tedious to write down explicit algebraic formulas for each of the b's, it is useful to note the following. The K slopes could just as well be computed by solving the K simultaneous linear equations

$$m_{11} b_1 + \cdots + m_{1K} b_K = m_{1y}$$

$$\cdots \qquad \cdots \qquad \cdots \qquad (2.35)$$

$$m_{K1} b_1 + \cdots + m_{KK} b_K = m_{Ky},$$

where the m's are "moments about the mean":

$$m_{jk} = \sum_{t=1}^{T} (x_{tj} - \bar{x}_j)(x_{tk} - \bar{x}_k) \qquad (j, k = 1, \ldots, K),$$

$$m_{jy} = \sum_{t=1}^{T} (x_{tj} - \bar{x}_j)(y_t - \bar{y}) \qquad (j = 1, \ldots, K).$$

Thereupon the intercept could be computed from

$$b_0 = (\sum y - b_1 \sum x_1 - \cdots - b_K \sum x_K)/T = \bar{y} - b_1\bar{x}_1 - \cdots - b_K\bar{x}_K.$$

(2.36)

These alternative formulas are of analytic interest. They imply that the slopes of a linear sample regression function do not depend upon the levels of the variables: Changing the levels by an additive constant affects the means (and hence the intercept) but not the moments about the means.

The calculated values of y, again denoted by $\hat{y}$, are

$$\hat{y}_t = b_0 + b_1 x_{t1} + \cdots + b_K x_{tK} \qquad (t = 1, \ldots, T),$$

(2.37)

and the residuals, again denoted by e, are

$$e_t = y_t - \hat{y}_t \qquad (t = 1, \ldots, T).$$

(2.38)

It is readily confirmed that

$$\sum e = \sum x_1 e = \cdots = \sum x_K e = 0,$$

(2.39)

whence

$$\sum \hat{y} = \sum (y - e) = \sum y,$$
$$\sum \hat{y}e = \sum (b_0 + b_1 x_1 + \cdots + b_K x_K)e = 0,$$

(2.40)

$$\sum y^2 = \sum (\hat{y} + e)^2 = \sum \hat{y}^2 + \sum e^2.$$

(2.41)

Adoption of a matrix formulation will permit us to write down a compact expression for the solution and will also be of use for analysis. First, we define a "dummy" variable x_0 that is identically equal to unity. Then (2.29) may be written $E(y \mid x_1, \ldots, x_K) = \beta_0 x_0 + \beta_1 x_1 + \cdots + \beta_K x_K$ or indeed as $E(y \mid \mathbf{x}') = \mathbf{x}'\boldsymbol{\beta}$, where $\mathbf{x}' = (x_0, x_1, \ldots, x_K)$ and $\boldsymbol{\beta}' = (\beta_0, \beta_1, \ldots, \beta_K)$. Then we define the matrices

$$\mathbf{y} = \begin{pmatrix} y_1 \\ \vdots \\ y_t \\ \vdots \\ y_T \end{pmatrix}, \quad \mathbf{X} = \begin{pmatrix} x_{10} & x_{11} & \cdots & x_{1K} \\ \vdots & \vdots & & \vdots \\ x_{t0} & x_{t1} & \cdots & x_{tK} \\ \vdots & \vdots & & \vdots \\ x_{T0} & x_{T1} & \cdots & x_{TK} \end{pmatrix}, \quad \mathbf{e} = \begin{pmatrix} e_1 \\ \vdots \\ e_t \\ \vdots \\ e_T \end{pmatrix}, \quad \mathbf{b} = \begin{pmatrix} b_0 \\ b_1 \\ \vdots \\ b_K \end{pmatrix}.$$

The T equations of (2.31) are then compactly expressed as

$$\mathbf{y} = \mathbf{Xb} + \mathbf{e}; \qquad (2.42)$$

the $1 + K$ conditions of (2.32) are

$$\mathbf{X'e} = \mathbf{0}; \qquad (2.43)$$

the $1 + K$ equations of (2.33) are

$$\mathbf{X'(y - Xb)} = \mathbf{0}; \qquad (2.44)$$

or those of (2.34) are

$$\mathbf{X'Xb} = \mathbf{X'y}. \qquad (2.45)$$

The solution of the system is

$$\mathbf{b} = \mathbf{(X'X)}^{-1}\mathbf{X'y} \qquad (2.46)$$

provided the inverse matrix exists—see Section 2.4, item 3, for more on this proviso. If we denote the $T \times 1$ vector of calculated values of y by $\hat{\mathbf{y}}$, then the T equations of (2.37) are

$$\hat{\mathbf{y}} = \mathbf{Xb} \qquad (2.47)$$

and those of (2.38) are

$$\mathbf{e} = \mathbf{y} - \hat{\mathbf{y}}. \qquad (2.48)$$

Similarly, the properties of (2.39) to (2.41) are expressed

$$\mathbf{X'e} = \mathbf{0}, \qquad (2.49)$$

$$\boldsymbol{\iota}'\hat{\mathbf{y}} = \boldsymbol{\iota}'\mathbf{y}, \qquad \hat{\mathbf{y}}'\mathbf{e} = \mathbf{b'X'e} = 0, \qquad (2.50)$$

where $\boldsymbol{\iota}' = (1 \cdots 1)$ is the $1 \times T$ vector of 1's, and

$$\mathbf{y'y} = (\hat{\mathbf{y}} + \mathbf{e})'(\hat{\mathbf{y}} + \mathbf{e}) = \hat{\mathbf{y}}'\hat{\mathbf{y}} + \mathbf{e'e}. \qquad (2.51)$$

Example 2.5

For the set of observations y_t, x_{t1}, x_{t2} ($t = 1, \ldots, 6$) given below, the calculation of the linear sample regression function is shown along with numerical checks of its properties. Note that y and x_1 are the same as the y and x of Example 2.4. A scatter diagram of these observations is given in Figure 2.2.

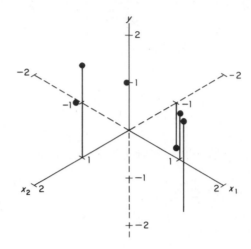

Figure 2.2   Scatter diagram of y on x_1 and x_2.

t	x_1	x_2	y
1	−1	0	0
2	0	1	2
3	1	0	1
4	2	1	2
5	0	−1	−1
6	0	0	1

$$T = 6 \qquad \sum x_1 = 2 \qquad \sum x_2 = 1 \qquad \sum y = 5$$

$$\sum x_1^2 = 6 \qquad \sum x_1 x_2 = 2 \qquad \sum x_1 y = 5$$

$$\sum x_2^2 = 3 \qquad \sum x_2 y = 5$$

$$\sum y^2 = 11$$

$$m_{11} = 32/6 \qquad m_{12} = 10/6 \qquad m_{1y} = 20/6$$

$$m_{22} = 17/6 \qquad m_{2y} = 25/6$$

$$m_{yy} = 41/6$$

[The m's are computable as, for example, $m_{1y} = (T\sum x_1 y - \sum x_1 \sum y)/T$].

$$(32/6)b_1 + (10/6)b_2 = 20/6$$
$$(10/6)b_1 + (17/6)b_2 = 25/6$$

$$b_1 = [(17/6)(20/6) - (10/6)(25/6)]/[(32/6)(17/6) - (10/6)^2]$$

$$= 15/74$$

$$b_2 = [(32/6)(25/6) - (10/6)(20/6)]/[(32/6)(17/6) - (10/6)^2]$$

$$= 100/74$$

$$b_0 = [5 - (15/74)(2) - (100/74)(1)]/6$$

$$= 40/74$$

x_1	x_2	y	$\hat{y}$	e	$x_1 e$	$x_2 e$	$\hat{y}e$	y^2	$\hat{y}^2$	e^2
-1	0	0	$25/74$	$-25/74$	$25/74$	$0/74$	$-625/74^2$	0	$625/74^2$	$625/74^2$
0	1	2	$140/74$	$8/74$	$0/74$	$8/74$	$1120/74^2$	4	$19600/74^2$	$64/74^2$
1	0	1	$55/74$	$19/74$	$19/74$	$0/74$	$1045/74^2$	1	$3025/74^2$	$361/74^2$
2	1	2	$170/74$	$-22/74$	$-44/74$	$-22/74$	$-3740/74^2$	4	$28900/74^2$	$484/74^2$
0	-1	-1	$-60/74$	$-14/74$	$0/74$	$14/74$	$840/74^2$	1	$3600/74^2$	$196/74^2$
0	0	1	$40/74$	$34/74$	$0/74$	$0/74$	$1360/74^2$	1	$1600/74^2$	$1156/74^2$
Sums			$370/74$	$0/74$	$0/74$	$0/74$	$0/74^2$	11	$57350/74^2$	$2886/74^2$
			$=5$	$=0$	$=0$	$=0$	$=0$	$=2442/222$	$=2325/222$	$=117/222$

2.4. SUPPLEMENTARY REMARKS

1. There are at least two other approaches which lead to obtaining the coefficients of a sample regression function from (2.34) or (2.45) — (2.12) in the simple case. The first, a merely descriptive approach, seeks the linear function which "fits the data best" in terms of the *least-squares criterion*. The second, a statistical inference approach, seeks the *minimum variance linear unbiased estimators* of the population parameters β's in a fully specified stochastic framework, the "classical linear regression model." These traditional approaches are developed in Johnston (1963, pp. 9–20, 106–114). The present development, although not embedded in a fully specified stochastic model, does view linear regression as a method for estimation of population parameters rather than as mere descriptive curve fitting. The population regression function is seen as a conditional expectation, and we seek the function whose sample properties are analogous to the population properties of the population regression function. As compared with the least-squares approach, the present development may be viewed as less arbitrary, since it has some inferential content. As compared with the classical linear regression model approach,

the present development perhaps has the virtue of suggesting that the linear sample regression function has attractive properties even when the strict set of classical assumptions is not met. At the very least, we have a fresh interpretation of the well-known computations.

Our view of regression analysis as an attempt to estimate a conditional expectation is in line with the arguments of Wold (1964). These arguments indicate that, as estimators of the coefficients of a conditional expectation, least-squares coefficients have desirable properties from the point of view of statistical inference under very general assumptions. Another stimulus of the present approach is the treatment in Klein (1962, pp. 33–38).

2. It is important to recognize that population regression functions are not always the proper objectives of empirical research. A large portion of the econometric literature is devoted to contexts—such as the "error-in-variables model" and the "simultaneous structural equation model"—in which the conditional expectation of y given the x's is not of direct interest. As a consequence, least-squares regression of y on the x's is not an appropriate estimation procedure. These models fall outside the scope of this book; the interested reader is directed to Johnston (1963, Chaps. 6 and 9) and Malinvaud (1966, Chaps. 10, 16–20).

3. It is important to recognize that the equations of (2.34) or (2.45) may not have a unique solution. In the case of a single conditioning variable discussed in Section 2.2, the necessary and sufficient condition for the existence of a unique solution is from (2.13) readily seen to be $\sum_{t=1}^{T} (x_t - \bar{x})^2 \neq 0$. Assuming this condition is the same as saying that x_t is not constant over the whole sample. The general case of K conditioning variables is best analyzed in matrix algebra terms. The unique solution (2.46) will exist if and only if the inverse matrix $(\mathbf{X}'\mathbf{X})^{-1}$ exists, that is, if and only if the rank of the $(1 + K) \times (1 + K)$ matrix $\mathbf{X}'\mathbf{X}$ is $1 + K$, that is, if and only if the $1 + K$ columns of $\mathbf{X}'\mathbf{X}$ are linearly independent. This is equivalent to the condition that the $T \times (1 + K)$ matrix $\mathbf{X}$ has rank $1 + K$, that is, that the $1 + K$ columns of $\mathbf{X}$ are linearly independent. To see the equivalence: If the columns of $\mathbf{X}$ are linearly dependent, then there exists a $(1 + K) \times 1$ vector $\mathbf{c} \neq \mathbf{0}$ such that $\mathbf{Xc} = \mathbf{0}$, in which case $\mathbf{X}'\mathbf{Xc} = \mathbf{X}'\mathbf{0} = \mathbf{0}$, so that the columns of $\mathbf{X}'\mathbf{X}$ are linearly dependent. Conversely, if the columns of $\mathbf{X}'\mathbf{X}$ are linearly dependent, then there exists a $(1 + K) \times 1$ vector $\mathbf{d} \neq \mathbf{0}$ such that $\mathbf{X}'\mathbf{Xd} = \mathbf{0}$, in which case $\mathbf{d}'\mathbf{X}'\mathbf{Xd} = \mathbf{d}'\mathbf{0} = 0$; but $\mathbf{d}'\mathbf{X}'\mathbf{Xd} = (\mathbf{Xd})'(\mathbf{Xd})$ is a sum of squares which can be zero only if $\mathbf{Xd} = \mathbf{0}$, in which case the columns of $\mathbf{X}$ are linearly dependent.

To repeat, if the columns of $\mathbf{X}$ are linearly dependent, then the conditions (2.43) do not suffice to determine a unique vector $\mathbf{b}$. Explicitly, if $\mathbf{Xc} = \mathbf{0}$

for some $\mathbf{c} \neq \mathbf{0}$, then if $\mathbf{b}$ is a solution to $\mathbf{X'Xb} = \mathbf{X'y}$ so is $\mathbf{a} = \mathbf{b} + \mathbf{c}$:
$\mathbf{X'Xa} = \mathbf{X'X(b + c)} = \mathbf{X'Xb} + \mathbf{X'Xc} = \mathbf{X'y} + \mathbf{X'0} = \mathbf{X'y}$. It is useful to
reexamine this result from the point of view of least-squares theory. To
do this, let us denote the columns of $\mathbf{X}$ as $\mathbf{X}_0, \mathbf{X}_1, \ldots, \mathbf{X}_K$; then the linear
sample regression function is

$$\hat{\mathbf{y}} = b_0 \mathbf{X}_0 + b_1 \mathbf{X}_1 + \cdots + b_K \mathbf{X}_K. \tag{2.52}$$

From least-squares theory we know that the b's are chosen to minimize the
sum of squared residuals $(\mathbf{y} - \hat{\mathbf{y}})'(\mathbf{y} - \hat{\mathbf{y}})$. That is, the b's are the coefficients
of the linear combination of the column vectors of $\mathbf{X}$ which comes closest
to $\mathbf{y}$ in the least-squares sense. If the columns of $\mathbf{X}$ are linearly dependent,
there is a set of c's not all of which are zero such that

$$\mathbf{0} = c_0 \mathbf{X}_0 + c_1 \mathbf{X}_1 + \cdots + c_K \mathbf{X}_K. \tag{2.53}$$

Then adding (2.52) and (2.53) gives

$$\hat{\mathbf{y}} = a_0 \mathbf{X}_0 + a_1 \mathbf{X}_1 + \cdots + a_K \mathbf{X}_K, \tag{2.54}$$

where $a_k = b_k + c_k$ $(k = 0, \ldots, K)$. Thus there is a distinct linear combination of the column vectors of $\mathbf{X}$—a different set of regression coefficients—which gives the very same $\hat{\mathbf{y}}$, and hence fits just as well.

In view of this we shall, until further notice, be assuming that $\mathbf{X}$ has full
column rank, so that the coefficients of the linear sample regression function are unique. In Chapter 6 we return to this question.

4. From the point of view of description of the sample relationship
between two variables y and x, the linear sample regression of y on x
developed in Section 2.2 is only one of many alternatives. Consider the
scatter diagram of the x, y observations with y measured on the vertical
axis and x on the horizontal axis. Our regression is characterized by the
fact that it is the straight line which minimizes the sum of squared deviations measured in the vertical direction. One might also consider the straight
line which minimizes the sum of squared deviations measured in the
horizontal direction; clearly this could be obtained by taking the linear
sample regression of x on y. A third alternative, to which we shall have
occasion to refer later, is the *orthogonal regression*, characterized by the
fact that it is the straight line which minimizes the sum of squared deviations measured in the perpendicular direction (that is, perpendicular to
the line).

In general, the lines obtained by these three alternative criteria do not coincide. For a comprehensive comparison, covering also the case of more than two variables, see Malinvaud (1966, pp. 7–11, 40–44). For our purposes, the following will suffice. Consider each of the lines expressed in terms of y as a function of x: $y = a + bx$. The slope b and intercept a are alternatively computed as follows:

Vertical minimization (y-on-x regression):

$$b_v = m_{xy}/m_{xx}, \qquad a_v = \bar{y} - b_v\bar{x}. \tag{2.55}$$

Horizontal minimization (x-on-y regression):

$$b_h = m_{yy}/m_{xy}, \qquad a_h = \bar{y} - b_h\bar{x}. \tag{2.56}$$

Perpendicular minimization (orthogonal regression):

$$b_p = 2m_{xy}/[(m_{xx} - m_{yy}) + \sqrt{(m_{xx} - m_{yy})^2 + 4m_{xy}^2},$$
$$a_p = \bar{y} - b_p\bar{x}. \tag{2.57}$$

In these formulas, the m's again denote moments about the mean: $m_{xx} = \sum(x - \bar{x})^2$, $m_{yy} = \sum(y - \bar{y})^2$, $m_{xy} = \sum(x - \bar{x})(y - \bar{y})$. The feature which may make the orthogonal regression attractive as a descriptive device in some contexts is that it provides a compromise between the two natural regressions. More specifically,

$$|b_v| \leq |b_p| \leq |b_h|; \tag{2.58}$$

the slope of the orthogonal regression always lies between the two other slopes. It will be instructive for the reader to compute all three regression lines for the data of Example 2.4 and to plot them on the same scatter diagram.

Chapter 3

Regression Coefficients

IN CHAPTER 2 WE OFFERED A HEURISTIC JUSTIFICATION for the fitting of linear functions by the method of least squares. We considered the case where the population regression function was linear:

$$E(y \mid x_1, \ldots, x_K) = \beta_0 + \beta_1 x_1 + \cdots + \beta_K x_K. \qquad (3.1)$$

We proposed to estimate this by fitting a linear function between the *regressand* y and the *regressors* $x_1, \ldots, x_K$:

$$\hat{y} = b_0 + b_1 x_1 + \cdots + b_K x_K, \qquad (3.2)$$

where the b's are chosen to make the sample *residuals* $e = y - \hat{y}$ have some of the properties in the sample that population disturbances $\varepsilon = y - E(y \mid x_1, \ldots, x_K)$ have in the population, namely, zero mean and un-correlatedness with the x's. We have noted that the sample regression function—henceforth denoted SRF—thus determined is a least-squares function: Of all possible linear functions of the regressors it has the minimum sum of squared (vertical) deviations in the sample. We have suggested but not yet established a serious justification of our proposal from the point of view of statistical inference.

In the present chapter we shall be concerned with some problems of interpretation of the coefficients of the SRF. These problems will be familiar to those who have used linear regression analysis in empirical research. To begin, we note that in (3.1) each β_k ($k = 1, \ldots, K$) is a partial derivative, measuring the effect of a unit change in a regressor upon the conditional expectation of the regressand with all other regressors held constant; thus $\beta_k = \partial E(y \mid x_1, \ldots, x_K) / \partial x_k$. Correspondingly, in (3.2) each b_k is an estimate of the effect upon the conditional expectation of y of a unit change in x_k with all other x's held constant. For this reason the b_k's are called *partial regression coefficients*, and it is sometimes said that a partial regression coefficient measures the effect of one regressor after controlling for the effects of other regressors. This is in contrast to the *gross* (= *simple*) *regression coefficient* obtained by regressing y on that one regressor alone, for in the simple regression of y on x_k alone, effects of the other regressors have obviously not been controlled for.

To properly understand this distinction we should investigate in general terms the relation between partial and gross regression coefficients. This investigation will also shed some light on what happens when we add or delete one or more regressors in an empirical problem. As an illustration of the kind of empirical situation to which our analysis is relevant, we cite some demand functions for tobacco estimated by Koutsoyianni-Kokkova (1962, Chap. 2):

$$\hat{Q} = 0.3184M, \tag{3.3a}$$

$$\hat{Q} = 0.2648M + 0.2321P, \tag{3.3b}$$

$$\hat{Q} = 0.3416M + 0.2935P - 0.1401L, \tag{3.3c}$$

where Q = tobacco consumption, M = money income, P = tobacco price, L = cost of living index, and the observations are annual logarithms for the United Kingdom 1950–1959. (We have omitted the intercept because it is irrelevant for our present purposes.) Note how the coefficients change as additional regressors are introduced; our analysis should help to explain what is happening.

3.2. PARTIAL AND GROSS REGRESSION COEFFICIENTS: TWO-REGRESSOR CASE

Consider the case of a linear regression of a regressand y on the two regressors x_1 and x_2:

$$y_t = b_0 + b_1 x_{t1} + b_2 x_{t2} + e_t \qquad (t = 1, \ldots, T). \tag{3.4}$$

It will be convenient to (1) assume that the variables are already measured as deviations about their sample means so that they now have zero mean and the intercept drops out, (2) adopt a more explicit notation, (3) drop the observation subscript for simplicity, and (4) henceforth use LSRF to denote linear sample regression function. Accordingly we rewrite (3.4) as

$$y = b_{y1 \cdot 2} x_1 + b_{y2 \cdot 1} x_2 + e_{y \cdot 12}, \tag{3.5}$$

where $b_{y1 \cdot 2}$ denotes the slope of y with respect to x_1 in the LSRF in which the other regressor is x_2, $b_{y2 \cdot 1}$ denotes the slope of y with respect to x_2 in the LSRF in which the other regressor is x_1, and $e_{y \cdot 12}$ denotes the residual in the LSRF of y on x_1 and x_2.

The equations (2.35) from which the slopes are computed are now written

$$m_{11} b_{y1 \cdot 2} + m_{12} b_{y2 \cdot 1} = m_{1y}, \tag{3.6}$$

$$m_{21} b_{y1 \cdot 2} + m_{22} b_{y2 \cdot 1} = m_{2y}. \tag{3.7}$$

Following a standard procedure for solving a pair of simultaneous linear equations we obtain

$$(m_{11} m_{22} - m_{12}^2) b_{y1 \cdot 2} = (m_{22} m_{1y} - m_{12} m_{2y}), \tag{3.8}$$

$$(m_{11} m_{22} - m_{12}^2) b_{y2 \cdot 1} = (m_{11} m_{2y} - m_{12} m_{1y}), \tag{3.9}$$

using $m_{21} = m_{12}$.

Now let us consider the LSRF of y on x_1 alone, which on our present notational scheme may be written

$$y = b_{y1} x_1 + e_{y \cdot 1}, \tag{3.10}$$

where b_{y1} denotes the slope of y on x_1 in the LSRF in which no other regressors are used, and $e_{y \cdot 1}$ denotes the residual in the LSRF of y on x_1 alone. The solution for the slope in (2.13) may be written

$$b_{y1} = m_{1y}/m_{11}. \tag{3.11}$$

Further, consider the LSRF of x_2 on x_1:

$$x_2 = b_{21} x_1 + x_{2 \cdot 1}, \tag{3.12}$$

where b_{21} denotes the slope of x_2 on x_1 in the *auxiliary* LSRF of x_2 on x_1 alone, and $x_{2 \cdot 1}$ denotes the residual in the auxiliary LSRF of x_2 on x_1. The regression (3.12) is called auxiliary because it is introduced only to assist in our investigation of (3.5). In any event its slope is computed as [compare (2.13)]

$$b_{21} = m_{12}/m_{11}. \tag{3.13}$$

If we insert (3.11) and (3.13) in (3.6), we find after rearrangement that

$$b_{y1 \cdot 2} = b_{y1} - b_{21} b_{y2 \cdot 1}. \tag{3.14}$$

This says that the partial regression coefficient of y on x_1 (in the presence of x_2) *equals* the gross regression coefficient of y on x_1 *minus* the (auxiliary) regression coefficient of x_2 on x_1 *times* the partial regression coefficient of y on x_2 (in the presence of x_1). The reader may find it instructive to check out (3.14) using the first two tobacco demand functions together with the fact that the slope of the LSRF of P on M was 0.2309.

If we turn (3.14) around into

$$b_{y1} = b_{y1 \cdot 2} + b_{21} b_{y2 \cdot 1}, \tag{3.15}$$

we see that if we regress y on x_1 alone, the slope we obtain will differ from the x_1 slope we would obtain if we regressed y on x_1 and x_2 together. Indeed it is equal to that slope plus the product of two terms: the slope of x_2 on x_1, and the x_2 slope in the regression of y on x_1 and x_2 together. As a memory aid to (3.15) we may use a familiar differential calculus formula:

If $y = f(x_1, x_2)$ and $x_2 = g(x_1)$,
then $dy/dx_1 = \partial y/\partial x_1 + (dx_2/dx_1)(\partial y/\partial x_2)$.

An analogue to the sample formula (3.15) may be found in the population regression function—henceforth PRF. Suppose that there is a linear population regression function—henceforth LPRF—of y on x_1 and x_2. In our present notational scheme we may then write

$$y = \beta_{y1 \cdot 2} x_1 + \beta_{y2 \cdot 1} x_2 + \varepsilon, \tag{3.16}$$

where the first two terms on the right give the conditional expectation $E(y \mid x_1, x_2)$. By the rules (2.21) and (2.22) on expectations, it follows that

$$E(y \mid x_1) = \beta_{y1 \cdot 2} x_1 + \beta_{y2 \cdot 1} E(x_2 \mid x_1) + E(\varepsilon \mid x_1). \tag{3.17}$$

Now $E(\varepsilon \mid x_1) = 0$ by (2.27) and the rule (2.25), so that if the auxiliary PRF of x_2 on x_1 were also linear, say,

$$E(x_2 \mid x_1) = \beta_{21} x_1, \tag{3.18}$$

then (3.17) would reduce to

$$E(y \mid x_1) = (\beta_{y1 \cdot 2} + \beta_{21}\beta_{y2 \cdot 1}) x_1. \tag{3.19}$$

The slope of this PRF has the same form as does the slope in the SRF, (3.15).

To suggest the practical applicability of (3.15) we cite the following, adapted from Griliches (1957). Suppose that we believe that the production function $Y = Y(L, K)$ is linear, where $Y = $ output, $L = $ labor, and $K = $ capital. We are interested in the marginal product of labor, $\partial Y / \partial L$; we may presume that $\partial Y / \partial L$ and also the marginal product of capital, $\partial Y / \partial K$, are positive. Our sample consists of a cross section of firms for which we observe Y and L, but unfortunately not K. We regress Y on L, obtaining the slope b_{YL}. What can we say about the L slope, $b_{YL \cdot K}$, that we could have found if we had had observations on K? It is plausible to suppose that the unobserved K was positively correlated with the observed L—that is, firms which used much L also used much K. If so, the auxiliary slope b_{KL}—which we cannot actually compute—would be positive. But then (3.15) tells us, at least, that the gross slope b_{YL} is an overestimate of the partial slope $b_{YL \cdot K}$; recall the presumption that both $b_{YL \cdot K}$ and $b_{YK \cdot L}$ are positive. Speaking somewhat loosely, we may say that in the regression of Y on L, labor receives credit for the effect of capital.

It should be recognized by this point that (3.14) and (3.15) provide an exact answer to the question of what happens to the slopes when we add or delete a regressor. To see the sense in which a partial regression coefficient measures the effect of one regressor after controlling for the effect of the other, we proceed as follows. Consider the residuals from the auxiliary regression (3.12), the $x_{2 \cdot 1}$. In a fairly natural sense *these residuals measure x_2 after allowance for the effects* of x_1. Now if we were to take the regression of y on these residuals, obtaining in a natural notation

$$y = b_{y(2 \cdot 1)} x_{2 \cdot 1} + e_{y \cdot (2 \cdot 1)}, \tag{3.20}$$

the slope in this *residual regression* would be identical with the x_2 slope in the original multiple regression, that is,

$$b_{y(2 \cdot 1)} = b_{y2 \cdot 1}. \tag{3.21}$$

For the equation determining the slope of (3.20) is [compare (2.13)]

$$m_{(2 \cdot 1)(2 \cdot 1)} b_{y(2 \cdot 1)} = m_{(2 \cdot 1)y} \qquad (3.22)$$

in obvious notation. But

$$
\begin{aligned}
m_{(2 \cdot 1)(2 \cdot 1)} &= \sum x_{2 \cdot 1}^2 = \sum (x_2 - b_{21} x_1)^2 \\
&= m_{22} + b_{21}^2 m_{11} - 2b_{21} m_{12} \\
&= m_{22} - m_{12}^2 / m_{11}, \qquad (3.23) \\
m_{(2 \cdot 1)y} &= \sum x_{2 \cdot 1} y = \sum (x_2 - b_{21} x_1) y \\
&= m_{2y} - b_{21} m_{1y} \\
&= m_{2y} - (m_{12} m_{1y}) / m_{11}, \qquad (3.24)
\end{aligned}
$$

since $b_{21} = m_{21} / m_{11}$. Inserting (3.23) and (3.24) in (3.22), and multiplying through by m_{11}, we see that (3.22), the equation determining $b_{y(2 \cdot 1)}$, is identical with (3.9), the equation determining $b_{y2 \cdot 1}$, so that (3.21) is indeed correct.

Thus (3.21) provides a precise meaning for the idea that a partial regression coefficient measures the effect of one regressor after controlling for the effect of the other. This interpretation is strengthened if we try to control for the effect of x_1 on y as well as on x_2. For, suppose we take the *double residual regression*:

$$e_{y \cdot 1} = b_{(y \cdot 1)(2 \cdot 1)} x_{2 \cdot 1} + e_{(y \cdot 1) \cdot (2 \cdot 1)} \qquad (3.25)$$

in natural notation. The equation determining its slope is

$$m_{(2 \cdot 1)(2 \cdot 1)} b_{(y \cdot 1)(2 \cdot 1)} = m_{(2 \cdot 1)(y \cdot 1)}. \qquad (3.26)$$

But

$$
\begin{aligned}
m_{(2 \cdot 1)(y \cdot 1)} &= \sum x_{2 \cdot 1} e_{y \cdot 1} = \sum (x_2 - b_{21} x_1)(y - b_{y1} x_1) \\
&= \sum (x_2 - b_{21} x_1) y = m_{(2 \cdot 1)y} \qquad (3.27)
\end{aligned}
$$

since $\sum (x_2 - b_{21} x_1)(b_{y1} x_1) = b_{y1}(m_{12} - b_{21} m_{11}) = 0$ in view of $b_{21} = m_{12}/m_{11}$. Thus (3.26) is in fact the same as (3.22). We conclude that the double residual regression slope is the same as the residual regression slope and hence that it is the same as the multiple regression slope:

$$b_{(y \cdot 1)(2 \cdot 1)} = b_{y2 \cdot 1}. \qquad (3.28)$$

But note that if we only control y for x_1 and not x_2 for x_1, we do not get the "right" answer. That is, consider the *stepwise regression*:

$$e_{y \cdot 1} = b_{(y \cdot 1)2} x_2 + e_{(y \cdot 1) \cdot 2}. \tag{3.29}$$

The equation determining its slope is

$$m_{22} b_{(y \cdot 1)2} = m_{2(y \cdot 1)}, \tag{3.30}$$

which, since $m_{2(y \cdot 1)} = \sum x_2(y - b_{y1}x_1) = m_{2y} - (m_{1y}/m_{11})m_{12}$, becomes after multiplication through by m_{11}:

$$m_{11}m_{22} b_{(y \cdot 1)2} = m_{11}m_{2y} - m_{12} m_{1y}. \tag{3.31}$$

This differs from (3.9). Indeed if we anticipate some results of Chapter 4 [(4.12) and (4.13)], we can show that the stepwise slope understates the multiple regression slope in absolute value. From (3.31) and (3.9) we have

$$b_{(y \cdot 1)2} = [(m_{11}m_{22} - m_{12}^2)/(m_{11}m_{22})]b_{y2 \cdot 1}$$
$$= (1 - R_{21}^2)b_{y2 \cdot 1}, \tag{3.32}$$

where $R_{21}^2 = m_{12}^2/m_{11}m_{22}$ is the coefficient of determination in the auxiliary regression of x_2 on x_1. Since any R^2 lies between 0 and 1 we see that $|b_{(y \cdot 1)2}| \le |b_{y2 \cdot 1}|$, with equality holding if and only if $R_{21}^2 = 0$ or $b_{y2 \cdot 1} = 0$.

Example 3.1

The results of this section may be illustrated with the data of Examples 2.4 and 2.5.

Auxiliary regression x_2 on x_1:

x_1	x_2				$\hat{x}_{2 \cdot 1}$	$x_{2 \cdot 1}$
−1	0	$T = 6$	$\sum x_1 = 2$	$\sum x_2 = 1$	−4/16	4/16
0	1				1/16	15/16
1	0		$\sum x_1^2 = 6$	$\sum x_1 x_2 = 2$	6/16	−6/16
2	1				11/16	5/16
0	−1		$m_{11} = 32/6$	$m_{12} = 10/6$	1/16	−17/16
0	0				1/16	−1/16

$$b_{21} = (10/6)/(32/6) = 5/16$$

$$\text{Intercept} = [1 - (5/16)(2)]/6 = 1/16$$

$$\text{Calculated value} = \hat{x}_{2 \cdot 1} = (1/16) + (5/16)x_1$$

Check of (3.14): $b_{y1} - b_{21} \quad b_{y2 \cdot 1} = b_{y1 \cdot 2}$

$$(5/8) - (5/16)(100/74) = 15/74$$

Residual regression y on $x_{2 \cdot 1}$:

$x_{2 \cdot 1}$	y
4/16	0
15/16	2
−6/16	1
5/16	2
−17/16	−1
−1/16	1

$T = 6 \quad \sum x_{2 \cdot 1} = 0 \qquad \sum y \quad = 5$

$\sum x_{2 \cdot 1}^2 = 592/16^2 \qquad \sum x_{2 \cdot 1}y = 50/16$

$m_{(2 \cdot 1)(2 \cdot 1)} = 37/16 \qquad m_{(2 \cdot 1)y} = 50/16$

$$b_{y(2 \cdot 1)} = (50/16)/(37/16) = 50/37$$

Check of (3.21): $b_{y(2 \cdot 1)} = b_{y2 \cdot 1}$

$$50/37 = 100/74$$

Double residual regression $e_{y \cdot 1}$ on $x_{2 \cdot 1}$:

$x_{2 \cdot 1}$	$e_{y \cdot 1}$
4/16	0/8
15/16	11/8
−6/16	−2/8
5/16	1/8
−17/16	−13/8
−1/16	3/8

$T = 6 \quad \sum x_{2 \cdot 1} = 0 \qquad \sum e_{y \cdot 1} \quad = 0$

$\sum x_{2 \cdot 1}^2 = 592/16^2 \qquad \sum x_{2 \cdot 1}e_{y \cdot 1} = 400/[(16)(8)]$

$m_{(2 \cdot 1)(2 \cdot 1)} = 37/16 \qquad m_{(2 \cdot 1)(y \cdot 1)} = 50/16$

$$b_{(y \cdot 1)(2 \cdot 1)} = (50/16)/(592/16^2) = 800/592$$

Check of (3.28): $b_{(y \cdot 1)(2 \cdot 1)} = b_{y2 \cdot 1}$

$$800/592 = 100/74$$

Stepwise regression $e_{y \cdot 1}$ on x_2:

x_2	$e_{y \cdot 1}$
0	0/8
1	11/8
0	−2/8
1	1/8
−1	−13/8
0	3/8

$T = 6$ $\sum x_2 = 1$ $\sum e_{y \cdot 1} = 0$

$\sum x_2^2 = 3$ $\sum x_2 e_{y \cdot 1} = 25/8$

$m_{22} = 17/6$ $m_{2(y \cdot 1)} = 25/8$

$$b_{(y \cdot 1)2} = (25/8)/(17/6) = 75/68$$

Check of (3.32): $b_{(y \cdot 1)2} = [1 - m_{12}^2/(m_{11}m_{22})]b_{y2 \cdot 1}$

$$75/68 = \{1 - (10/6)^2/[(32/6)(17/6)]\}(100/74)$$

3.3. PARTIAL AND GROSS REGRESSION COEFFICIENTS: K-REGRESSOR CASE

The results of Section 3.2 have close parallels when there are more than two regressors. The parallels are emphasized when matrix algebra is used.

Consider then the case of linear regression of a regressand y on a set of K regressors $x_1, \ldots, x_K$. Assuming that the variables are already measured as deviations about their sample means so that they now have zero mean and the intercept drops out, and adopting a more explicit notation, we rewrite (2.42) as

$$\mathbf{y} = \mathbf{X}_1 \mathbf{b}_{y1 \cdot 2} + \mathbf{X}_2 \mathbf{b}_{y2 \cdot 1} + \mathbf{e}_{y \cdot 12}, \qquad (3.33)$$

where $\mathbf{y}$ is the $T \times 1$ vector of regressand observations; $\mathbf{X}_1$ is the $T \times H$ matrix of observations on the regressors $x_1, \ldots, x_H$, where $1 \leq H < K$; $\mathbf{b}_{y1 \cdot 2}$ is the $H \times 1$ vector of slopes of y with respect to $x_1, \ldots, x_H$ in the LSRF in which the other regressors are $x_{H+1}, \ldots, x_K$; $\mathbf{X}_2$ is the $T \times (K - H)$ matrix of observations on the regressors $x_{H+1}, \ldots, x_K$; $\mathbf{b}_{y2 \cdot 1}$ is the $(K - H) \times 1$ vector of slopes of y with respect to $x_{H+1}, \ldots, x_K$ in the LSRF in which the other regressors are $x_1, \ldots, x_H$; and $\mathbf{e}_{y \cdot 12}$ is the $T \times 1$ vector of residuals from the LSRF of y on $x_1, \ldots, x_H, x_{H+1}, \ldots, x_K$.

Consider the normal equation system (2.45) which determines the slopes. In partitioned form it may be written

$$\begin{pmatrix} \mathbf{X}_1' \mathbf{X}_1 & \mathbf{X}_1' \mathbf{X}_2 \\ \mathbf{X}_2' \mathbf{X}_1 & \mathbf{X}_2' \mathbf{X}_2 \end{pmatrix} \begin{pmatrix} \mathbf{b}_{y1 \cdot 2} \\ \mathbf{b}_{y2 \cdot 1} \end{pmatrix} = \begin{pmatrix} \mathbf{X}_1' \mathbf{y} \\ \mathbf{X}_2' \mathbf{y} \end{pmatrix};$$

that is,

$$X_1'X_1 b_{y1 \cdot 2} + X_1'X_2 b_{y2 \cdot 1} = X_1'y, \tag{3.34}$$

$$X_2'X_1 b_{y1 \cdot 2} + X_2'X_2 b_{y2 \cdot 1} = X_2'y. \tag{3.35}$$

[For a discussion of matrix partitioning, the reader may consult Johnston (1963, pp. 88–90) or Goldberger (1964, pp. 12–15).] Multiplying (3.35) through by $X_1'X_2(X_2'X_2)^{-1}$ and subtracting from (3.34) gives

$$[X_1'X_1 - X_1'X_2(X_2'X_2)^{-1}X_2'X_1]b_{y1 \cdot 2} = [X_1'y - X_1'X_2(X_2'X_2)^{-1}X_2'y].$$
$$\tag{3.36}$$

Similarly, multiplying (3.34) through by $X_2'X_1(X_1'X_1)^{-1}$ and subtracting from (3.35) gives

$$[X_2'X_2 - X_2'X_1(X_1'X_1)^{-1}X_1'X_2]b_{y2 \cdot 1} = [X_2'y - X_2'X_1(X_1'X_1)^{-1}X_1'y].$$
$$\tag{3.37}$$

Equations (3.36) and (3.37) determine $b_{y1 \cdot 2}$ and $b_{y2 \cdot 1}$, respectively.

Now let us consider the LSRF of y on the set $x_1, \ldots, x_H$, which in our present notation is

$$y = X_1 b_{y1} + e_{y \cdot 1}, \tag{3.38}$$

where b_{y1} is the $H \times 1$ vector of slopes of y on $x_1, \ldots, x_H$ in the LSRF in which no other regressors are used, and $e_{y \cdot 1}$ is the $T \times 1$ vector of residuals in the LSRF of y on $x_1, \ldots, x_H$ alone. The solution for the slope may be written [compare (2.46)]

$$b_{y1} = (X_1'X_1)^{-1}X_1'y. \tag{3.39}$$

Further, consider the *set of $K - H$ auxiliary LSRF's* of each variable in the set $x_{H+1}, \ldots, x_K$ upon the set $x_1, \ldots, x_H$:

$$X_2 = X_1 B_{21} + X_{2 \cdot 1}, \tag{3.40}$$

where B_{21} is a $H \times (K - H)$ *matrix*, each *column* of which is a $H \times 1$ vector of slopes in an auxiliary regression of a variable in X_2 on X_1 in which no other regressors are used, and $X_{2 \cdot 1}$ is a $T \times (K - H)$ *matrix*, each *column* of which is a $T \times 1$ vector of residuals in an auxiliary regression. Since each column of $\mathbf{B}_{21}$ is a regression slope vector, the entire matrix may be written

$$\mathbf{B}_{21} = (\mathbf{X}_1'\mathbf{X}_1)^{-1}\mathbf{X}_1'\mathbf{X}_2 . \tag{3.41}$$

Let us insert (3.39) and (3.41) in (3.34). We find after rearrangement that

$$\mathbf{b}_{y1 \cdot 2} = \mathbf{b}_{y1} - \mathbf{B}_{21}\mathbf{b}_{y2 \cdot 1} . \tag{3.42}$$

The typical row of (3.42) says that the partial regression coefficient of $\mathbf{y}$ on a variable in $\mathbf{X}_1$ (in the presence of $\mathbf{X}_2$ and the rest of $\mathbf{X}_1$) *equals* the partial regression coefficient of $\mathbf{y}$ on that variable (in the absence of $\mathbf{X}_2$) *minus* the sum of partial auxiliary regression coefficients of each of the variables in $\mathbf{X}_2$ on $\mathbf{X}_1$ *times* the partial regression coefficients of $\mathbf{y}$ on those variables (in the presence of $\mathbf{X}_1$ and the rest of $\mathbf{X}_2$).

If we turn (3.42) around into

$$\mathbf{b}_{y1} = \mathbf{b}_{y1 \cdot 2} + \mathbf{B}_{21}\mathbf{b}_{y2 \cdot 1} \tag{3.43}$$

we see that if we regress $\mathbf{y}$ on $\mathbf{X}_1$ alone, the slope vector we obtain will differ from the $\mathbf{X}_1$ slope vector we would get if we regressed $\mathbf{y}$ on $\mathbf{X}_1$ and $\mathbf{X}_2$ together. Indeed it is equal to that slope vector plus the product of two terms: the auxiliary slope matrix of $\mathbf{X}_2$ on $\mathbf{X}_1$, and the $\mathbf{X}_2$ slope vector in the regression of $\mathbf{y}$ on $\mathbf{X}_1$ and $\mathbf{X}_2$ together.

It should be recognized that (3.42) and (3.43) provide exact answers to the questions of what happens to the slopes if we add or delete one or more regressors. To see the sense in which one partial slope vector controls for the effect of the other variables, we proceed as follows. Consider the residuals from the auxiliary regressions (3.40), namely, the $\mathbf{X}_{2 \cdot 1}$. In a fairly natural sense *these residuals measure* $\mathbf{X}_2$ *after allowance for the effects of* $\mathbf{X}_1$. If we were to take the regression of $\mathbf{y}$ on these residuals, obtaining

$$\mathbf{y} = \mathbf{X}_{2 \cdot 1}\mathbf{b}_{y(2 \cdot 1)} + \mathbf{e}_{y \cdot (2 \cdot 1)} \tag{3.44}$$

in obvious notation, then the slope vector in this *residual regression* would be identical with the $\mathbf{X}_2$ slope vector in the original full multiple regression:

$$\mathbf{b}_{y(2 \cdot 1)} = \mathbf{b}_{y2 \cdot 1} . \tag{3.45}$$

For the equation determining the slope of (3.44) is

$$\mathbf{X}_{2\cdot1}'\mathbf{X}_{2\cdot1}\mathbf{b}_{y(2\cdot1)} = \mathbf{X}_{2\cdot1}'\mathbf{y}. \tag{3.46}$$

But

$$\mathbf{X}_{2\cdot1}'\mathbf{X}_{2\cdot1} = (\mathbf{X}_2 - \mathbf{X}_1\mathbf{B}_{21})'(\mathbf{X}_2 - \mathbf{X}_1\mathbf{B}_{21})$$

$$= \mathbf{X}_2'\mathbf{X}_2 + \mathbf{B}_{21}'\mathbf{X}_1'\mathbf{X}_1\mathbf{B}_{21} - \mathbf{B}_{21}'\mathbf{X}_1'\mathbf{X}_2 - \mathbf{X}_2'\mathbf{X}_1\mathbf{B}_{21}$$

$$= \mathbf{X}_2'\mathbf{X}_2 - \mathbf{X}_2'\mathbf{X}_1(\mathbf{X}_1'\mathbf{X}_1)^{-1}\mathbf{X}_1'\mathbf{X}_2, \tag{3.47}$$

$$\mathbf{X}_{2\cdot1}'\mathbf{y} = (\mathbf{X}_2 - \mathbf{X}_1\mathbf{B}_{21})'\mathbf{y} = \mathbf{X}_2'\mathbf{y} - \mathbf{B}_{21}'\mathbf{X}_1'\mathbf{y}$$

$$= \mathbf{X}_2'\mathbf{y} - \mathbf{X}_2'\mathbf{X}_1(\mathbf{X}_1'\mathbf{X}_1)^{-1}\mathbf{X}_1'\mathbf{y}, \tag{3.48}$$

since

$$\mathbf{B}_{21} = (\mathbf{X}_1'\mathbf{X}_1)^{-1}\mathbf{X}_1'\mathbf{X}_2.$$

Inserting (3.47) and (3.48) in (3.46) we see that (3.46), the equation defining $\mathbf{b}_{y(2\cdot1)}$, is identical with (3.37), the equation defining $\mathbf{b}_{y2\cdot1}$, so that (3.45) is indeed correct.

Thus (3.45) gives precise meaning to the idea that a partial regression coefficient vector measures the effects of one subset of regressors after controlling for effects of the other subset. This interpretation is strengthened if we try to control for the effect of $\mathbf{X}_1$ on $\mathbf{y}$ as well as on $\mathbf{X}_2$. For, suppose we take the *double residual regression*

$$\mathbf{e}_{y\cdot1} = \mathbf{X}_{2\cdot1}\mathbf{b}_{(y\cdot1)(2\cdot1)} + \mathbf{e}_{(y\cdot1)\cdot(2\cdot1)}. \tag{3.49}$$

The equation determining its slope vector is

$$\mathbf{X}_{2\cdot1}'\mathbf{X}_{2\cdot1}\mathbf{b}_{(y\cdot1)(2\cdot1)} = \mathbf{X}_{2\cdot1}'\mathbf{e}_{y\cdot1}. \tag{3.50}$$

Now

$$\mathbf{X}_{2\cdot1}'\mathbf{e}_{y\cdot1} = (\mathbf{X}_2 - \mathbf{X}_1\mathbf{B}_{21})'(\mathbf{y} - \mathbf{X}_1\mathbf{b}_{y1})$$

$$= \mathbf{X}_2'\mathbf{y} - \mathbf{X}_2'\mathbf{X}_1(\mathbf{X}_1'\mathbf{X}_1)^{-1}\mathbf{X}_1'\mathbf{y}, \tag{3.51}$$

since

$$\mathbf{B}_{21}'\mathbf{X}_1'\mathbf{X}_1\mathbf{b}_{y1} = \mathbf{X}_2'\mathbf{X}_1(\mathbf{X}_1'\mathbf{X}_1)^{-1}\mathbf{X}_1'\mathbf{X}_1(\mathbf{X}_1'\mathbf{X}_1)^{-1}\mathbf{X}_1'\mathbf{y}$$

$$= \mathbf{X}_2'\mathbf{X}_1(\mathbf{X}_1'\mathbf{X}_1)^{-1}\mathbf{X}_1'\mathbf{y} = \mathbf{B}_{21}'\mathbf{X}_1'\mathbf{y} = \mathbf{X}_2'\mathbf{X}_1\mathbf{b}_{y1}.$$

Thus (3.50) is in fact the same as (3.46). We conclude that the double residual regression slope vector is the same as the residual regression slope vector and hence is the same as the multiple regression slope subvector:

$$\mathbf{b}_{(y\cdot 1)(2\cdot 1)} = \mathbf{b}_{y2\cdot 1}. \tag{3.52}$$

But note that if we only control $\mathbf{y}$ for $\mathbf{X}_1$ and not $\mathbf{X}_2$ for $\mathbf{X}_1$, we do not get the "right" answer. That is, consider the *stepwise regression*

$$\mathbf{e}_{y\cdot 1} = \mathbf{X}_2\,\mathbf{b}_{(y\cdot 1)2} + \mathbf{e}_{(y\cdot 1)\cdot 2}. \tag{3.53}$$

The equation determining its slope vector is

$$\mathbf{X}_2'\mathbf{X}_2\,\mathbf{b}_{(y\cdot 1)2} = \mathbf{X}_2'\,\mathbf{e}_{y\cdot 1}, \tag{3.54}$$

which, since $\mathbf{X}_2'\,\mathbf{e}_{y\cdot 1} = \mathbf{X}_2'(\mathbf{y} - \mathbf{X}_1\mathbf{b}_{y1}) = \mathbf{X}_2'[\mathbf{y} - \mathbf{X}_1(\mathbf{X}_1'\mathbf{X}_1)^{-1}\mathbf{X}_1'\mathbf{y}]$, becomes

$$\mathbf{X}_2'\mathbf{X}_2\,\mathbf{b}_{(y\cdot 1)2} = [\mathbf{X}_2'\,\mathbf{y} - \mathbf{X}_2'\,\mathbf{X}_1(\mathbf{X}_1'\mathbf{X}_1)^{-1}\mathbf{X}_1'\mathbf{y}]. \tag{3.55}$$

Comparing this with (3.37), we see that

$$\mathbf{b}_{(y\cdot 1)2} = [\mathbf{I} - (\mathbf{X}_2'\,\mathbf{X}_2)^{-1}\mathbf{X}_2'\,\mathbf{X}_1(\mathbf{X}_1'\mathbf{X}_1)^{-1}\mathbf{X}_1'\mathbf{X}_2]\mathbf{b}_{y2\cdot 1}. \tag{3.56}$$

While this expression does not have a simple intuitive interpretation in the general case, it does indicate that the stepwise slope vector differs from the multiple regression slope subvector. For an interpretation in the case in which $\mathbf{X}_2$ contains only one variable, see Goldberger (1964, p. 195).

3.4. SUPPLEMENTARY REMARKS

1. The algebraic connections between partial and gross regression coefficients have, of course, long been known, and the literature contains many formulas for computing the effect of adding or dropping a regressor. The recent revival of interest in this subject is perhaps attributable to the analysis of "specification error" developed by Theil (1957) and to the critique of stepwise regression offered by Goldberger and Jochems (1961). Another influence has been the development of computer programs for quickly trying out different sets of regressions. A critical evaluation of such procedures is given by Draper and Smith (1966, Chap. 6). (*Caution*: Draper and Smith use the term "stagewise" where we use "stepwise.")

2. The key role played by the auxiliary regression slope in the preceding discussion should be emphasized. In the two-regressor case suppose that x_1 and x_2 are uncorrelated in the sample, that is, $m_{12} = 0$. Then $b_{21} = 0$; the LSRF of x_2 on x_1 is constant. It follows immediately that $b_{y1} = b_{y1 \cdot 2}$ and $b_{y2} = b_{y2 \cdot 1}$ (where b_{y2} is the gross regression coefficient of y on x_2), so that the slopes of the multiple regression could just as well be obtained by two separate simple regressions. Also $b_{(y \cdot 1)2} = b_{y2 \cdot 1}$, so that stepwise regression gives the "right" answer. Similarly, in the K-regressor case, suppose that each variable in $\mathbf{X}_1$ is uncorrelated with every variable in $\mathbf{X}_2$, that is, $\mathbf{X}_1' \mathbf{X}_2 = \mathbf{0}$ (recall that the variables are taken to have zero means). Then $\mathbf{B}_{21} = \mathbf{0}$; the LSRF of each variable in $\mathbf{X}_2$ upon the set of variables in $\mathbf{X}_1$ is constant. It follows immediately that $\mathbf{b}_{y1} = \mathbf{b}_{y1 \cdot 2}$ and $\mathbf{b}_{y2} = \mathbf{b}_{y2 \cdot 1}$ (where $\mathbf{b}_{y2}$ is the slope vector in the LSRF of y on $\mathbf{X}_2$ alone), so that the slope vector of the full multiple regression could just as well be obtained by two separate "shorter" multiple regressions. Also $\mathbf{b}_{(y \cdot 1)2} = \mathbf{b}_{y2 \cdot 1}$, so that stepwise regression gives the "right" answer. By an extension of this argument, we see that if each regressor is uncorrelated in the sample with every other regressor, the $K \times 1$ slope vector could just as well be obtained by K separate simple regressions. The computational and analytical advantages of such uncorrelatedness among the regressors are clear. Unfortunately, since the economic researcher typically takes the data as they come, and does not have control over the correlation among the x's, he is unlikely to possess samples with this attractive feature.

Chapter 4

Coefficients of Determination

4.1. INTRODUCTION

I<small>N</small> C<small>HAPTER</small> 2 <small>WE OBTAINED A RELATION AMONG</small> certain sums of squares which are associated with linear regression. This relation was

$$\sum y^2 = \sum \hat{y}^2 + \sum e^2, \qquad (4.1)$$

where $\sum y^2$ is the sum of squares of the observed values of y, $\sum \hat{y}^2$ is the sum of squares of their calculated values, and $\sum e^2$ is the sum of squares of the residuals. Another version turns out to be more useful. Since $\sum y = \sum \hat{y}$, the equality in (4.1) is preserved if we subtract $(\sum y)^2/T$ from the left side and $(\sum \hat{y})^2/T$ from the right side. After rearrangement the result is

$$\sum (y - \bar{y})^2 = \sum (y - \bar{y})^2 + \sum e^2. \qquad (4.2)$$

In (4.2) the sum of squared deviations of the observed y about their mean (henceforth *total sum of squares*, or SST) is given as the sum of two terms: the sum of squared deviations of the calculated y about their mean (henceforth *regression sum of squares*, or SSR) and the sum of squared deviations of the residuals about their (zero) mean (henceforth *error sum of squares*, or SSE).

In a fairly natural sense, SSR measures the sample variation of the regressand that is accounted for by (=attributable to = explained by) the regressors—that is, by the fitted linear function of the regressors. In the same way, SSE measures the sample variation of the regressand which is not accounted for by (=not attributable to = not explained by) the regressors—that is, by the fitted linear function of the regressors. The two components together add up to SST, which measures the sample variation of the regressand.

It seems natural, and it is certainly customary, to take as a measure of the "goodness-of-fit" of the LSRF the so-called *coefficient of determination*,

$$R^2 = \sum (\hat{y} - \bar{y})^2 / \sum (y - \bar{y})^2 = 1 - \sum e^2 / \sum (y - \bar{y})^2, \qquad (4.3)$$

which gives the proportion of the sample variation of the regressand which is accounted for by (or attributable to, or explained by) the regressors—that is, by the fitted linear function of the regressors. Clearly our LSRF, which is chosen to minimize the sum of squared residuals, also maximizes the proportion explained.

Now on the approach we have been taking, the coefficient of determination may be considered as an indicator of the strength of the stochastic relationship between y and the x's. For, in principle, we should say that the relationship between y and the x's is strong if the conditional distribution of y changes substantially as the x's take on different values. But in practice we concentrate only on the means of these conditional distributions, so that we may say that the relationship between y and the x's is strong if the conditional expectation of y changes substantially as the x's take on different values. To give precision to the idea of changing substantially it seems natural to relate the variance of the conditional expectation of y to the variance of y itself; that is, to consider as the population measure of strength the population parameter

$$P^2 = \{E[E(y \mid x') - E(y)]^2\} / \{E[y - E(y)]^2\}. \qquad (4.4)$$

It is apparent that the coefficient of determination R^2 is simply the sample analogue of P^2.

In the present chapter we shall be concerned with some problems of interpretation of the coefficient of determination. These problems will be familiar to those who have used linear regression analysis in empirical research. To begin, suppose that we have fitted the LSRF

$$\hat{y} = b_0 + b_1 x_1 + \cdots + b_K x_K \qquad (4.5)$$

and computed the R^2. This *multiple coefficient of determination* measures the proportion of the variation of y accounted for by all the x's taken together. How is this related to the *simple coefficients of determination*, obtained by regressing y on each of the x's separately? Those simple, or gross, R^2's measure the proportion of the variation of y accounted for by each of the x's taken separately. Further, is it possible to allocate the multiple coefficient of determination among the regressors so as to be able to say that this much of the explanation is due to x_1, this much to x_2, and so forth?

To answer these questions we should investigate in general terms the relation between simple and multiple coefficients of determination. This investigation will also shed some light on what happens when we add or delete one or more regressors in an empirical problem. As an illustration of the kind of empirical problem to which our analysis is relevant, we may note the multiple coefficients of determination for the three tobacco demand functions of Koutsoyianni-Kokkova cited in (3.3); they are, respectively, 0.964, 0.970, and 0.973. Note that R^2 changes as additional regressors are introduced; our analysis should help to explain what is happening.

It will come as no surprise to learn that the analysis of Chapter 3 is of great help in our present investigation. Before proceeding, however, it will be useful to note several algebraic facts.

First, let $R^2_{y \cdot 1, \ldots, K}$ denote the coefficient of determination in the multiple regression of y on $x_1, \ldots, x_K$; let $\hat{y}$ denote the calculated values in that regression; and let $R^2_{y\hat{y}}$ denote the coefficient of determination in the simple regression of y on $\hat{y}$. It is not hard to show that $R^2_{y\hat{y}} = R^2_{y \cdot 1, \ldots, K}$; the argument is as follows. When we regress y on $\hat{y}$, we find the slope

$$b_{y\hat{y}} = \sum [(\hat{y} - \bar{\hat{y}})(y - \bar{y})] / \sum (\hat{y} - \bar{\hat{y}})^2 \qquad (4.6)$$

and the intercept

$$a_{y\hat{y}} = \bar{y} - b_{y\hat{y}} \bar{\hat{y}}. \qquad (4.7)$$

Multiplying out the right side of (4.6) and drawing upon (2.38) to (2.40) for $\sum \hat{y}y = \sum \hat{y}(\hat{y} + e) = \sum \hat{y}^2$ and $\sum \hat{y} = \sum y$, we find that $b_{y\hat{y}} = 1$, and thus from (4.7) that $a_{y\hat{y}} = 0$. This means that the calculated values in the regression of y on $\hat{y}$ are the $\hat{y}$ themselves, so that the residuals from this regression are just the residuals from the multiple regression. The two regressions have the same regressand and the same residuals; therefore, they have the same SST and SSE; therefore, they have the same coefficient of determination.

Next, from (2.37) to (2.40) we have

$$\sum \hat{y}^2 = \sum \hat{y}(\hat{y} + e) = \sum \hat{y}y = \sum (b_0 + b_1 x_1 + \cdots + b_K x_K)y$$
$$= b_0 \sum y + b_1 \sum x_1 y + \cdots + b_K \sum x_K y. \tag{4.8}$$

After inserting (2.36) and rearranging, this is

$$\sum (\hat{y} - \bar{y})^2 = b_1 m_{1y} + \cdots + b_K m_{Ky} = \sum_{k=1}^{K} b_k m_{ky}. \tag{4.9}$$

Since $m_{yy} = \sum (y - \bar{y})^2$, we may express the coefficient of determination of y on $x_1, \ldots, x_K$ as

$$R_{y \cdot 1, \ldots, K}^2 = \sum_{k=1}^{K} b_k m_{ky}/m_{yy}. \tag{4.10}$$

It we specialize to the case $K = 1$, we obtain

$$\sum (\hat{y} - \bar{y})^2 = b_1 m_{1y}, \tag{4.11}$$

and then a familiar expression for the simple coefficient of determination of the regressand y on a single regressor:

$$R_{y1}^2 = b_1 m_{1y}/m_{yy} = m_{1y}^2/m_{11} m_{yy}. \tag{4.12}$$

Finally, from (4.3) we see that any coefficient of determination is the ratio of two sums of squares in which the numerator cannot exceed the denominator, so that

$$0 \le R^2 \le 1. \tag{4.13}$$

4.2. MULTIPLE AND SIMPLE COEFFICIENTS OF DETERMINATION: TWO-REGRESSOR CASE

Returning to the main theme of this chapter, we now take up the case of a linear regression of a regressand y on the two regressors x_1 and x_2 discussed in Section 3.2:

$$y = b_{y1 \cdot 2} x_1 + b_{y2 \cdot 1} x_2 + e_{y \cdot 12}. \tag{4.14}$$

Note that we again consider the variables as measured in terms of deviations about their sample means so that they have zero mean and the intercept drops out.

Let us write $\hat{y}_{.12}$ for the calculated value in this regression; thus

$$\hat{y}_{.12} = b_{y1.2}x_1 + b_{y2.1}x_2 = y - e_{y.12}. \tag{4.15}$$

These calculated values will have zero mean under the present convention, so that, from (4.9), we have for the SSR

$$\sum \hat{y}_{.12}^2 = b_{y1.2}m_{1y} + b_{y2.1}m_{2y}. \tag{4.16}$$

This may be written

$$\begin{aligned}
\sum \hat{y}_{.12}^2 &= (b_{y1} - b_{21}b_{y2.1})m_{1y} + b_{y2.1}m_{2y} \\
&= b_{y1}m_{1y} + b_{y2.1}(m_{2y} - b_{21}m_{1y}) \\
&= b_{y1}m_{1y} + b_{y2.1}m_{(2.1)y} \\
&= b_{y1}m_{1y} + b_{y(2.1)}m_{(2.1)y},
\end{aligned} \tag{4.17}$$

using in turn (3.14), the line above (3.24), and (3.21). Now the SSR in the regression of y on x_1 alone is $\sum \hat{y}_{.1}^2 = b_{y1}m_{1y}$, while the SSR in the residual regression of y on $x_{2.1}$ is $\sum \hat{y}_{.(2.1)}^2 = b_{y(2.1)}m_{(2.1)y}$. Thus

$$\sum \hat{y}_{.12}^2 = \sum \hat{y}_{.1}^2 + \sum \hat{y}_{.(2.1)}^2; \tag{4.18}$$

the SSR in the multiple regression of y on x_1 and x_2 together may be expressed as the sum of two terms: the SSR in the simple regression of y on x_1 alone, and the SSR in the residual regression of y on $x_{2.1}$.

Dividing (4.18) through by SST, which under the present convention is $\sum y^2$, we find

$$\begin{aligned}
R_{y.12}^2 &= \sum \hat{y}_{.12}^2 / \sum y^2 = \sum \hat{y}_{.1}^2 / \sum y^2 + \sum \hat{y}_{.(2.1)}^2 / \sum y^2 \\
&= R_{y.1}^2 + \sum \hat{y}_{.(2.1)}^2 / \sum y^2.
\end{aligned} \tag{4.19}$$

Thus the multiple coefficient of determination of y on x_1 and x_2 together may be expressed as the sum of two terms: the simple coefficient of determination of y on x_1 alone, and a nonnegative quantity (sums of squares cannot be negative). Thus adding an additional regressor cannot decrease the R^2.

Further, we have seen just above (3.28) that the slope of the double residual regression is identical with the slope of the residual regression. With all variables having zero means, it follows that the calculated values from the double residual regression, the $\hat{y}_{.1(2\cdot1)}$, are identical with the calculated values from the residual regression, the $\hat{y}_{.(2\cdot1)}$. Therefore, the second term on the right side of (4.19) can be written

$$\sum \hat{y}^2_{.(2\cdot1)}/\sum y^2 = \sum \hat{y}^2_{.1(2\cdot1)}/\sum y^2 = \left(\sum \hat{y}^2_{.1(2\cdot1)}/\sum e^2_{y\cdot1}\right)/\left(\sum e^2_{y\cdot1}/\sum y^2\right).$$

But it will be recognized that $\sum \hat{y}^2_{.1(2\cdot1)}$ is the SSR and $\sum e^2_{y\cdot1}$ is the SST in the double residual regression, so that their ratio is $R^2_{(y\cdot1)(2\cdot1)}$, the coefficient of determination in that regression. Also $\sum e^2_{y\cdot1}/\sum y^2$ is $(1 - R^2_{y\cdot1})$, 1 minus the coefficient of determination in the simple regression. Using these expressions we see that (4.19) may be written

$$R^2_{y\cdot12} = R^2_{y\cdot1} + (1 - R^2_{y\cdot1})R^2_{(y\cdot1)(2\cdot1)}. \qquad (4.20)$$

This says that the proportion of the variation in y explained by x_1 and x_2 together *equals* the sum of two terms: the proportion of the variation in y explained by x_1 alone, and the proportion of the variation in y unexplained by x_1 alone *times* the proportion of this unexplained variation which is explained by x_2 after controlling for x_1.

It should be recognized by this point that (4.20) provides an exact answer to the question of what happens to the coefficient of determination when we add or delete a regressor. The statistic $R^2_{(y\cdot1)(2\cdot1)}$ is sometimes called the *partial coefficient of determination* of y on x_2, given x_1, since it measures the proportion of y explained by x_2 after controlling both variables for x_1.

It is tempting to use (4.20) to allocate the multiple R^2 between the two regressors: saying that the first term on the right side measures x_1's share, and the second term on the right side measures x_2's share, of their joint explanation of y. But this temptation should be resisted. After all, if we reverse the roles of x_1 and x_2 throughout, we would come up with

$$R^2_{y\cdot12} = R^2_{y\cdot2} + (1 - R^2_{y\cdot2})R^2_{(y\cdot2)(1\cdot2)}. \qquad (4.21)$$

This says that the proportion of the variation in y explained by x_1 and x_2 together *equals* the sum of two terms: the proportion of the variation in y explained by x_2 alone, and the proportion of the variation in y unexplained by x_2 alone *times* the proportion of this unexplained variation which is explained by x_1 after controlling for x_2. In (4.21) it might seem that the first term on the right measures x_2's share, and the second term on the right measures x_1's share, of their joint explanation of y. Unfortunately, in general (4.20) and (4.21) give conflicting allocations of the multiple R^2 between the two regressors. We conclude that there is in general no unique way to allocate the joint explanation between the two explaining variables.

This conclusion is reinforced when we recognize that the multiple coefficient of determination is not equal to the sum of the two simple coefficients of determination:

$$R^2_{y \cdot 12} \neq R^2_{y \cdot 1} + R^2_{y \cdot 2} \quad \text{in general.} \quad (4.22)$$

For the equality to hold, we see from (4.18) that we must have $\sum \hat{y}^2_{\cdot (2 \cdot 1)} = \sum \hat{y}^2_{\cdot 2}$, but this is generally not the case. Incidentally the validity of (4.22) will be apparent to anyone who has attempted a multiple regression using two regressors, each of which has a simple R^2 of 0.90 with the regressand; surely the multiple R^2 was not 1.80. Nor is it true in general that $R^2_{y \cdot 12} < R^2_{y \cdot 1} + R^2_{y \cdot 2}$; a counterexample is given in Watts (1965) along with some other instructive material on the connections among simple, multiple, and partial coefficients of determination.

Example 4.1

The results of this section may be illustrated with the data of Examples 2.4, 2.5, and 3.1.

$$m_{11} = 32/6 \qquad m_{12} = 10/6 \qquad m_{1y} = 20/6$$
$$m_{22} = 17/6 \qquad m_{2y} = 25/6$$
$$m_{yy} = 41/6$$

$$b_{y1} = 5/8$$
$$b_{y1 \cdot 2} = 15/74 \qquad b_{y2 \cdot 1} = 100/74$$
$$m_{(2 \cdot 1)(2 \cdot 1)} = 37/16 \qquad m_{(2 \cdot 1)y} = 50/16 = m_{(2 \cdot 1)(y \cdot 1)}$$
$$m_{(y \cdot 1)(y \cdot 1)} = 304/64$$

Check of (4.11) and (4.12):

$$\sum (\hat{y}_{\cdot 1} - \bar{\hat{y}}_{\cdot 1})^2 = 400/64 - 5^2/6 = 100/48 = (5/8)(20/6)$$
$$= b_{y1} m_{1y}$$
$$R^2_{y \cdot 1} = \sum (\hat{y}_{\cdot 1} - \bar{\hat{y}}_{\cdot 1})^2 / m_{yy} = (100/48)/(41/6) = 25/82$$
$$= (20/6)^2/(32/6)(41/6) = m^2_{1y}/(m_{11} m_{yy})$$

Check of (4.16):

$$\sum (\hat{y}_{.12} - \bar{\hat{y}}_{.12})^2 = 2325/222 - 5^2/6 = 1400/222$$

$$= (15/74)(20/6) + (100/74)(25/6)$$

$$= b_{y1.2}m_{1y} \quad + b_{y2.1}m_{2y}$$

$$R_{y.12}^2 = \sum (\hat{y}_{.12} - \bar{\hat{y}}_{.12})^2/m_{yy} = (1400/222)/(41/6) = 1400/1517$$

$$\sum (\hat{y}_{.(2.1)} - \bar{\hat{y}}_{.(2.1)})^2 = b_{y(2.1)}m_{(2.1)y} = (100/74)(50/16)$$

$$= 625/148$$

Check of (4.18):

$$\sum (\hat{y}_{.12} - \bar{\hat{y}}_{.12})^2 = 1400/222 = 100/48 + 625/148$$

$$= \sum (\hat{y}_{.1} - \bar{\hat{y}}_{.1})^2 + \sum (\hat{y}_{.(2.1)} - \bar{\hat{y}}_{.(2.1)})^2$$

$$R_{(y.1)(2.1)}^2 = m_{(2.1)(y.1)}^2/[m_{(y.1)(y.1)}m_{(2.1)(2.1)}]$$

$$= (50/16)^2/[(304/64)(37/16)] = 625/703$$

Check of (4.20):

$$R_{y.12}^2 = 1400/1517 = 25/82 + (1 - 25/82)(625/703)$$

$$= R_{y.1}^2 + (1 - R_{y.1}^2)R_{(y.1)(2.1}^2$$

Check of (4.22):

$$R_{y.2}^2 = m_{2y}^2/(m_{22}m_{yy}) = (25/6)^2/[(17/6)(41/6)] = 625/697$$

$$R_{y.1}^2 + R_{y.2}^2 = 25/82 + 625/697 = 61,975/51,578$$

$$\neq 47,600/51,578 = 1400/1517 = R_{y.12}^2$$

4.3. MULTIPLE AND SIMPLE COEFFICIENTS OF DETERMINATION: K-REGRESSOR CASE

The results of Section 4.2 have close parallels when there are more than two regressors. The parallels are emphasized when matrix algebra is used.

Consider then the case of linear regression of a regressand y on a set of K regressors $x_1, \ldots, x_K$ discussed in Section 3.3:

$$\mathbf{y} = \mathbf{X}_1\mathbf{b}_{y1.2} + \mathbf{X}_2\mathbf{b}_{y2.1} + \mathbf{e}_{y.12}. \tag{4.23}$$

Note that we continue to consider the variables as measured in terms of deviations about their sample mean so that they have zero mean and the intercept drops out. Note also that the subscripts 1 and 2 now refer to sets of variables.

Let us write $\hat{y}_{.12}$ for the calculated value in this regression; thus

$$\hat{y}_{.12} = X_1 b_{y1 \cdot 2} + X_2 b_{y2 \cdot 1} = y - e_{y \cdot 12}. \tag{4.24}$$

These calculated values will have zero mean under the present convention so that the SSR is

$$\hat{y}'_{.12} \hat{y}_{.12} = b'_{y1 \cdot 2} X'_1 y + b'_{y2 \cdot 1} X'_2 y. \tag{4.25}$$

This may be written

$$
\begin{aligned}
\hat{y}'_{.12} \hat{y}_{.12} &= (b_{y1} - B_{21} b_{y2 \cdot 1})' X'_1 y + b'_{y2 \cdot 1} X'_2 y \\
&= b'_{y1} X'_1 y + b'_{y2 \cdot 1} (X'_2 y - B'_{21} X'_1 y) \\
&= b'_{y1} X'_1 y + b'_{y2 \cdot 1} X'_{2 \cdot 1} y \\
&= b'_{y1} X'_1 y + b'_{y(2 \cdot 1)} X'_{2 \cdot 1} y, \tag{4.26}
\end{aligned}
$$

using in turn (3.42), the line above (3.48), and (3.45). Now, the SSR in the regression of y on X_1 alone is $b'_{y1} X'_1 y$, while the SSR in the residual regression of y on $X_{2 \cdot 1}$ is $\hat{y}'_{.(2 \cdot 1)} \hat{y}_{.(2 \cdot 1)} = b'_{y(2 \cdot 1)} X'_{2 \cdot 1} y$. Thus

$$\hat{y}'_{.12} \hat{y}_{.12} = \hat{y}'_{.1} \hat{y}_{.1} + \hat{y}'_{.(2 \cdot 1)} \hat{y}_{.(2 \cdot 1)}. \tag{4.27}$$

The SSR in the multiple regression of y on X_1 and X_2 together equals the sum of two terms: the SSR in the regression of y on X_1 alone, and the SSR in the residual regression of y on $X_{2 \cdot 1}$.

Dividing (4.27) through by SST, which under the present convention is $y'y$, we find

$$
\begin{aligned}
R^2_{y \cdot 12} &= (\hat{y}'_{.12} \hat{y}_{.12})/(y'y) = (\hat{y}'_{.1} \hat{y}_{.1})/(y'y) + (\hat{y}'_{.(2 \cdot 1)} \hat{y}_{.(2 \cdot 1)})/(y'y) \\
&= R^2_{y \cdot 1} + (\hat{y}'_{.(2 \cdot 1)} \hat{y}_{.(2 \cdot 1)})/(y'y). \tag{4.28}
\end{aligned}
$$

Thus the multiple coefficient of determination of y on X_1 and X_2 together may be expressed as the sum of two terms: the coefficient of determination of y on X_1 alone, and a nonnegative quantity. Adding an additional set of regressors cannot decrease the R^2.

Further we have seen just above (3.52) that the slope of the double residual regression is identical with the slope of the residual regression. With all variables having zero means, it follows that the vector of calculated values from the double residual regression, the $\hat{y}_{\cdot 1(2 \cdot 1)}$, is identical with the vector of calculated values from the residual regression, $\hat{y}_{\cdot (2 \cdot 1)}$. By an argument parallel to that used in Section 4.2, we may then rewrite (4.28) as

$$R^2_{y \cdot 12} = R^2_{y \cdot 1} + (1 - R^2_{y \cdot 1})R^2_{(y \cdot 1)(2 \cdot 1)}. \qquad (4.29)$$

This says that the proportion of the variation in y explained by X_1 and X_2 together *equals* the sum of two terms: the proportion of the variation in y explained by X_1 alone, and the proportion of the variation in y unexplained by X_1 alone *times* the proportion of this unexplained variation which is explained by X_2 after controlling for X_1.

It should be recognized that (4.29) provides an exact answer to the question of what happens to the coefficient of determination when we add or delete a set of regressors. Once again we could have reversed the roles of X_1 and X_2 throughout and come up with

$$R^2_{y \cdot 12} = R^2_{y \cdot 2} + (1 - R^2_{y \cdot 2})R^2_{(y \cdot 2)(1 \cdot 2)}. \qquad (4.30)$$

In view of this and the fact that

$$R^2_{y \cdot 12} \neq R^2_{y \cdot 1} + R^2_{y \cdot 2} \qquad \text{in general,} \qquad (4.31)$$

we must conclude that there is in general no unique way of allocating the joint explanation between the two subsets of explaining variables.

4.4. SUPPLEMENTARY REMARKS

1. The algebraic connections between simple, multiple, and partial coefficients of determination have long been known. For further discussion of these and related coefficients see Ezekiel and Fox (1959, Chap. 12) and Johnston (1963, pp. 52–62).

2. The key role in the preceding analysis played by the auxiliary regression should be pointed out. Thus, in the two-regressor case, suppose that x_1 and x_2 are uncorrelated in the sample, so that $m_{12} = 0$, $b_{21} = 0$, and $b_{y(2 \cdot 1)} = b_{y2}$. We see from the line above (3.24) that $m_{(2 \cdot 1)y} = m_{2y}$

then also, whence from (4.17), $\sum \hat{y}^2_{\cdot 12} = \sum \hat{y}^2_{\cdot 1} + \sum \hat{y}^2_{\cdot 2}$, so that $R^2_{y \cdot 12} = R^2_{y \cdot 1} + R^2_{y \cdot 2}$. Thus the multiple coefficient of determination is the sum of the two simple coefficients of determination. Since (4.20) and (4.21) will also reduce to this form, there will be a unique way to allocate the joint explanation between the two explaining variables. Similar specializations apply in the K-regressor case when $\mathbf{X}'_1\mathbf{X}_2 = \mathbf{0}$.

3. Among the incidental results we have obtained is that SSR cannot decrease when additional regressors are used, or, to put it another way, that SSE cannot decrease when regressors are deleted. This conclusion can be derived in a considerably more elegant fashion. In fitting the LSRF of y on $x_1, \ldots, x_K$ we choose the linear function to minimize the sum of squared residuals. Suppose we delete the last $K - H$ regressors, and refit the regression. The resulting LSRF is the linear function of $x_1, \ldots, x_H$ which minimizes the sum of squared residuals. It can be described as the linear function of $x_1, \ldots, x_K$ which, subject to the constraint that $x_{H+1}, \ldots, x_K$ have zero coefficients, minimizes the sum of squared residuals. Clearly, a constrained minimum cannot be less than an unconstrained minimum; SSE cannot decrease.

Chapter 5

Precision of Estimation

5.1. INTRODUCTION

IN OUR APPROACH THE LSRF IS CHOSEN SO AS TO have its residuals uncorrelated with the regressors in the sample. As we have noted, this function is also a least-squares function and a maximum correlation function: Of all possible linear functions of the regressors $y^* = \sum_{k=0}^{K} b_k^* x_k$, our function $\hat{y} = \sum_{k=0}^{K} b_k x_k$ has the smallest sum of squared residuals from, and hence the maximum correlation with, the y's in the sample. In this traditional sense our LSRF is the "best fitting" line in the sample. But having recognized this, it seems natural to consider the following kinds of questions: How *much* better fitting is our line than alternative lines? How much would the sum of squared residuals rise, the coefficient of determination fall, the correlation between residuals and regressors deviate from zero, if we were to take some other line? How sensitive are $\sum e^2$, R^2, and $\sum x_k e$ to changes in the b's? How much different can we make the b's without changing $\sum e^2$, R^2, and $\sum x_k e$ much?

It is perhaps clear that an analysis of such questions can shed some light on the amount of precision which attaches to the individual regression coefficients. If the measures of goodness of fit are sensitive to changes in the regression coefficients, then considerable precision would attach to the regression coefficients. Usually we are interested in the individual regression coefficients, so that it is useful to have some idea of their reliability. While clear-cut answers to the question of reliability require a formal framework of statistical inference, it may be worthwhile to continue to postpone the formalities and to proceed along heuristic lines. Fortunately, it turns out that our heuristic measures of imprecision are intimately related to the classical measures that are based on formal statistical inference.

At this stage we may suggest the kind of empirical situation for which our discussion may be relevant. Suppose that we have estimated an LSRF and are surprised at the sign or magnitude of one of the slopes because it differs from what we would have expected on the basis of prior knowledge, experience, or preconception. It is then tempting to ask how much worse than the best line we would do in terms of goodness of fit if we imposed our preconceptions. How inconsistent with the data are our prior beliefs? Must we give them up because of this sample?

5.2. INDEX ERRORS OF REGRESSION COEFFICIENTS: ONE-REGRESSOR CASE

We start with the one-regressor case discussed in Section 2.2, taking the variables to be measured as deviations about their sample means as at (3.10) and (3.11). The LSRF is $\hat{y} = b_{y1}x$, with $b_{y1} = m_{1y}/m_{11}$ being the least-squares slope. Suppose that instead of the LSRF we take the function $y^* = (b_{y1} + \Delta_1)x$, where Δ_1 is the deviation of our arbitrary "pseudoslope" from the least-squares slope. Then the residuals from our "pseudofunction" are

$$e^* = y - y^* = y - (b_{y1} + \Delta_1)x = (y - b_{y1}x) - \Delta_1 x = e - \Delta_1 x, \quad (5.1)$$

where e denotes the residuals from the least-squares function. The sum of squared residuals from the pseudofunction will be

$$\sum e^{*2} = \sum (e - \Delta_1 x)^2 = \sum e^2 + \Delta_1^2 m_{11}, \qquad (5.2)$$

where $\sum e^2$ is the sum of squared residuals from the least-squares function, and we have used the fact that $\sum xe = 0$ by the condition determining the least-squares function. Incidentally, (5.2) confirms the least-squares property of b_{y1}, since $\sum e^{*2}$ clearly reaches a minimum at $\Delta_1 = 0$.

Now, according to (5.2), when m_{11} is small, that is, when the sample variation of the regressor x_1 is small, large deviations from the least-squares slope will not worsen the fit much. (Anyone who has tried to fit a freehand line to a scatter diagram in which the points are narrowly dispersed horizontally should appreciate the difficulty in choosing among alternative slopes in such a situation.) To be more specific, (5.2) says that we can take Δ_1 as large in absolute value as $\sqrt{1/m_{11}}$—that is, deviate from the least-squares slope by as much as $\sqrt{1/m_{11}}$ up or down—without increasing the sum of squared residuals by more than 1. If, for the present, we take an increase of 1 in the residual sum of squares as a standard, it then seems natural to adopt $\sqrt{1/m_{11}}$ as an indicator of the imprecision which attaches to the least-squares slope b_{y1}. When this indicator is large, that is, when m_{11} is small, we might attach little reliability to the least-squares slope. After all, quite different slopes do almost as well, in a specific goodness-of-fit sense, as it does.

Let us, then, define the *index error* Δ_k^* of a regression coefficient b_k as the maximum amount in absolute value by which that coefficient can be changed from its least-squares value without increasing the sum of squared residuals by more than 1. The argument just above has shown that in the present one-regressor case the index error of b_{y1} is $\Delta_1^* = \sqrt{1/m_{11}}$.

A geometric interpretation may be helpful. In Figure 5.1 we plot $(\sum e^{*2} - \sum e^2)$—that is, $\Delta_1^2 m_{11}$—as a function of Δ_1. This gives a parabola, opening upward, symmetrical about the vertical axis, with vertex at the origin. The parabola will be wide-mouthed—that is, even large (absolute) values of Δ_1 will give small values of $(\sum e^{*2} - \sum e^2)$—if m_{11} is small (that is, if $\sqrt{1/m_{11}}$ is large). The ordinate corresponding to the abscissas $\pm\Delta_1^* = \pm\sqrt{1/m_{11}}$ is 1. Our proposal is to take the interval $(-\Delta_1^*, \Delta_1^*)$, or rather its semilength Δ_1^*, as the measure of imprecision for the regression slope b_{y1}.

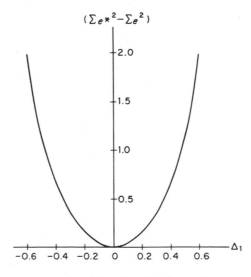

$$(\sum e^{*2} - \sum e^2)$$

Figure 5.1 $(\sum e^{*2} - \sum e^2) = \Delta_1^2 m_{11}$.

This type of analysis could be undertaken with respect to the coefficient of determination and the residual-regressor correlation. If we define $R^{*2} = 1 - \sum e^{*2}/\sum y^2$, then we have

$$R^2 - R^{*2} = (1 - \sum e^2/\sum y^2) - (1 - \sum e^{*2}/\sum y^2)$$
$$= -(\sum e^2 - \sum e^{*2})/\sum y^2$$
$$= \Delta_1^2 m_{11}/m_{yy} \qquad\qquad (5.3)$$

using (5.2). According to (5.3), $0.1\sqrt{m_{yy}/m_{11}}$ is the maximum amount by which one can deviate from the least-squares slope without reducing R^2 by more than 0.01. For the residual-regressor correlation, we have

$$R_{e^*x}^2 = (\sum xe^*)^2/(\sum x^2 \sum e^{*2}) = \Delta_1^2 m_{11}/\sum e^{*2}$$
$$= (\Delta_1^2 m_{11})/(\sum e^2 + \Delta_1^2 m_{11}), \qquad\qquad (5.4)$$

using $\sum xe = 0$ [whence $\sum xe^* = \sum x(e - \Delta_1 x) = -\Delta_1 m_{11}$], and (5.2). According to (5.4), given $\sum e^2$, if m_{11} is small, then large deviations from the least-squares slope will not induce much correlation between the residuals and the regressor. [A deviation of one index error induces a squared correlation coefficient of $1/(1 + \sum e^2)$.]

Example 5.1

The results of this section may be illustrated with the data of Example 2.4, which incidentally were used to construct Figure 5.1. The index error of b_{y1} is $\Delta_1^* = \sqrt{1/m_{11}} = \sqrt{6/32}$. The pseudofunction we consider is $y^* = 7/12 + 9/12\ x$. Since its intercept and slope satisfy $7/12 = \bar{y}^* - 9/12\ \bar{x}$, it may be viewed as if the variables had been expressed as deviations about sample means: This ensures that the results of this section are applicable.

x	y	y^*	e^*	e^{*2}	xe^*
-1	0	$-2/12$	$2/12$	$4/144$	$-2/12$
0	2	$7/12$	$17/12$	$289/144$	$0/12$
1	1	$16/12$	$-4/12$	$16/144$	$-4/12$
2	2	$25/12$	$-1/12$	$1/144$	$-2/12$
0	-1	$7/12$	$-19/12$	$361/144$	$0/12$
0	1	$7/12$	$5/12$	$25/144$	$0/12$
Sums		$60/12$	$0/12$	$696/144$	$-8/12$
		$= 5$	$= 0$	$= 58/12$	$= -2/3$

$\Delta_1 = 9/12 - 5/8 = 1/8$

$m_{11} = 32/6$

$\sum e^{*2} = 58/12 = 57/12 + 1/12 = 57/12 + (1/8)^2(32/6) = \sum e^2 + \Delta_1^2 m_{11}$

$\sum xe^* = -2/3 = -(1/8)(32/6) = -\Delta_1 m_{11}$

5.3. INDEX ERRORS OF REGRESSION COEFFICIENTS: TWO-REGRESSOR CASE

We proceed to the two-regressor case discussed in Section 3.2, taking the variables to be measured as deviations about their sample means as in (3.4) and (3.5). The LSRF is $\hat{y}_{.12} = b_{y1\cdot2}x_1 + b_{y2\cdot1}x_2$, where $b_{y1\cdot2}$ and $b_{y2\cdot1}$ are the least-squares slopes defined in (3.8) and (3.9). Suppose that instead of the LSRF we take the function $y^* = (b_{y1\cdot2} + \Delta_1)x_1 + (b_{y2\cdot1} + \Delta_2)x_2$, where Δ_1 and Δ_2 are the respective deviations of the pseudoslopes from the least-squares slopes. Then the residuals from this pseudofunction are

$$e^* = y - y^* = y - (b_{y1\cdot2} + \Delta_1)x_1 - (b_{y2\cdot1} + \Delta_2)x_2$$
$$= e - \Delta_1 x_1 - \Delta_2 x_2, \tag{5.5}$$

where $e = y - b_{y1\cdot2}x_1 - b_{y2\cdot1}x_2$ denotes the residuals from the least-squares function. The sum of squared residuals from the pseudofunction will be

$$\sum e^{*2} = \sum (e - \Delta_1 x_1 - \Delta_2 x_2)^2$$
$$= \sum e^2 + (\Delta_1^2 m_{11} + \Delta_2^2 m_{22} + 2\Delta_1\Delta_2 m_{12}), \tag{5.6}$$

where $\sum e^2$ is the sum of squared residuals from the least-squares function, and we have used the facts that $\sum x_1 e = \sum x_2 e = 0$ by the conditions determining the least-squares function. The least-squares property of $b_{y1\cdot2}$ and $b_{y2\cdot1}$ is confirmed by (5.6), since $\sum e^{*2}$ reaches a minimum at $\Delta_1 = \Delta_2 = 0$. [The expression in parentheses in (5.6) is a positive definite quadratic form; see Allen (1956, pp. 472–479) for a discussion of that concept.]

Again let us ask by how much we can deviate from the least-squares slopes without increasing the sum of squared residuals by more than 1. The answer is complicated by the presence of two slopes and even more by the presence of the cross-product term m_{12}.

Therefore, we shall first suppose that the regressors are uncorrelated in the sample so that $m_{12} = 0$ and (5.6) reduces to

$$\sum e^{*2} - \sum e^2 = \Delta_1^2 m_{11} + \Delta_2^2 m_{22}. \tag{5.7}$$

It is clear from (5.7) that in the present case the maximum (absolute) deviation of the first slope from its least-squares value consistent with $\left(\sum e^{*2} - \sum e^2\right)$ being less than or equal to 1 is $\sqrt{1/m_{11}}$. Further, it is clear that such a deviation must be accompanied by no deviation of the second slope from its least-squares value if $\left(\sum e^{*2} - \sum e^2\right)$ is not to exceed 1. Similarly, the maximum (absolute) deviation of the second slope from its least-squares value consistent with $\left(\sum e^{*2} - \sum e^2\right)$ being less than or equal to 1 is $\sqrt{1/m_{22}}$, and such a deviation must be accompanied by no deviation of the first slope from its least-squares value if $\left(\sum e^{*2} - \sum e^2\right)$ is not to exceed 1.

If we recall the definition of the index error of a regression coefficient, we see that in the present special case ($K = 2$, $m_{12} = 0$), the index error of $b_{y1 \cdot 2}$ is $\Delta_1^* = \sqrt{1/m_{11}}$ and the index error of $b_{y2 \cdot 1}$ is $\Delta_2^* = \sqrt{1/m_{22}}$. These are to be taken as measures of imprecision; when the index error is large—here when the sample variation of a regressor is small—the slope is imprecise, because quite different values do almost as well as it does.

It is worth noting that not only the index error points $\pm(\Delta_1^*, 0)$ and $\pm(0, \Delta_2^*)$ raise the residual sum of squares by exactly 1. Indeed, it is readily confirmed by substitution in (5.7) that, where p and q are any numbers such that $p^2 + q^2 = 1$, the combination $(p\Delta_1^*, q\Delta_2^*)$ makes $\left(\sum e^{*2} - \sum e^2\right) = 1$.

Again a geometric interpretation may be helpful. In Figure 5.2 we plot $\left(\sum e^{*2} - \sum e^2\right)$—that is, $\Delta_1^2 m_{11} + \Delta_2^2 m_{22}$—as a function of Δ_1 and Δ_2. This is an elliptic paraboloid—a bowl-shaped figure—opening upward, symmetric about the vertical axis, with vertex at the origin. Projecting onto the Δ_1, Δ_2 plane, as in Figure 5.3, we see that the contours of the elliptic paraboloid are a family of homothetic (that is, identically shaped and oriented) ellipses. The ellipses are concentric about the origin, and their axes coincide with the coordinate axes. Let us focus our attention on the *index ellipse*, that is, the contour for $\left(\sum e^{*2} - \sum e^2\right) = 1$. It is the locus of p, q combinations mentioned above. Its vertices are at the index error points $\pm(\Delta_1^*, 0)$ and $\pm(0, \Delta_2^*)$. The semilengths of its axes are Δ_1^* and Δ_2^* and it is these which we are proposing as measures of the imprecision of the respective least-squares slopes. These features of the index ellipse follow from elementary analytic geometry: $\Delta_1^2 m_{11} + \Delta_2^2 m_{22} = \Delta_1^2/(\sqrt{1/m_{11}})^2 + \Delta_2^2/(\sqrt{1/m_{22}})^2 = 1$ will be recognized as the equation of

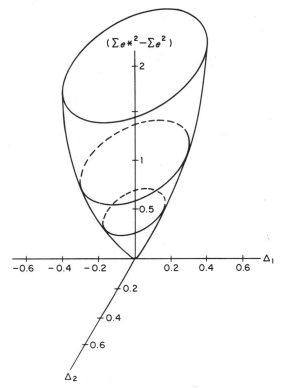

Figure 5.2 $(\sum e^{*2} - \sum e^2) = \Delta_1^2 m_{11} + \Delta_2^2 m_{22}$.

an ellipse with the properties mentioned. Further, since the area of such an ellipse is $\pi(\sqrt{1/m_{11}})(\sqrt{1/m_{22}})$, one might take $(\sqrt{1/m_{11}})(\sqrt{1/m_{22}}) = 1/\sqrt{m_{11}m_{22}}$ as a single measure of the joint imprecision of the least-squares slopes.

Having treated this very special case, where $m_{12} = 0$, we now allow the two regressors to be correlated in the sample so that we must return to the general expression of (5.6). By how much can we deviate from one least-squares slope without making $(\sum e^{*2} - \sum e^2)$ exceed 1? Consider the first slope. Certainly $\Delta_1 = \pm\sqrt{1/m_{11}}$ is still feasible since, accompanied by $\Delta_2 = 0$, it makes the parenthetical expression of (5.6), $(\Delta_1^2 m_{11} + \Delta_2^2 m_{22} + 2\Delta_1\Delta_2 m_{12})$, exactly equal to 1. But the presence of the cross-product term introduces the possibility of *compensating deviation* so that still larger (absolute) values of Δ_1 become feasible.

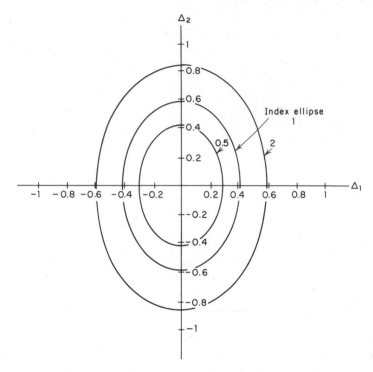

Figure 5.3   *Contours of* $(\sum e^{*2} - \sum e^2) = \Delta_1^2 m_{11} + \Delta_2^2 m_{22}$.

To find the index error of the first slope, we may maximize Δ_1^2 subject to the constraint $(\sum e^{*2} - \sum e^2) = 1$. We form the Lagrangian expression

$$L = \Delta_1^2 - \mu(\Delta_1^2 m_{11} + \Delta_2^2 m_{22} + 2\Delta_1\Delta_2 m_{12} - 1)$$

and differentiate it:

$$\partial L/\partial\Delta_1 = 2\Delta_1 - 2\mu(\Delta_1 m_{11} + \Delta_2 m_{12}), \qquad (5.8)$$

$$\partial L/\partial\Delta_2 = \quad - 2\mu(\Delta_1 m_{12} + \Delta_2 m_{22}), \qquad (5.9)$$

$$\partial L/\partial\mu = -(\Delta_1^2 m_{11} + \Delta_2^2 m_{22} + 2\Delta_1\Delta_2 m_{12} - 1). \qquad (5.10)$$

Setting (5.9) at zero we find

$$\Delta_2 = (-m_{12}/m_{22})\Delta_1 \qquad (5.11)$$

because the constraint is clearly binding, so that $\mu \neq 0$. Setting (5.10) at zero, inserting (5.11), and solving we get

$$\Delta_1^2 = m_{22}/(m_{11}m_{22} - m_{12}^2). \tag{5.12}$$

It can be confirmed that this locates a maximum. We conclude that in the general two-regressor case the index error of $b_{y1 \cdot 2}$ is $\Delta_1^* = \sqrt{m_{22}/(m_{11}m_{22} - m_{12}^2)}$. This is the largest (absolute) deviation of the first slope from its least-squares value which is consistent with no more than a unit increase in the sum of squared residuals. This deviation must be accompanied by the compensating deviation of the second slope defined by inserting Δ_1^* in (5.11), namely, $\Delta_{2(1*)} = (-m_{12}/m_{22})\Delta_1^*$, if no more than a unit increase is to occur.

Alternative derivations of the index error are perhaps instructive. If we treat the constraint as a quadratic equation in Δ_2,

$$m_{22}\Delta_2^2 + 2\Delta_1 m_{12}\Delta_2 + (\Delta_1^2 m_{11} - 1) = 0, \tag{5.13}$$

the discriminant of this equation is $(2\Delta_1 m_{12})^2 - 4m_{22}(\Delta_1^2 m_{11} - 1)$. For a quadratic equation to have real roots the discriminant must be non-negative, which amounts to requiring that

$$m_{22} - \Delta_1^2(m_{11}m_{22} - m_{12}^2) \geq 0. \tag{5.14}$$

Since the term in parentheses in (5.14) is nonnegative—it equals $m_{11}m_{22}(1 - R_{21}^2)$—it follows that the largest value of Δ_1^2 which satisfies the inequality is the one that makes it an equality, namely, $\Delta_1^2 = m_{22}/(m_{11}m_{22} - m_{12}^2)$. Thus our Δ_1^* is the largest (absolute) deviation of the first slope from its least-squares value for which compensation is feasible.

A third derivation proceeds by reducing the present two-regressor case to the one-regressor case treated in Section 5.2. By an argument parallel to that underlying (3.28), the first slope in the multiple regression, namely, $b_{y1 \cdot 2}$, is identical with the slope in the double residual regression of $e_{y \cdot 2}$ on $x_{1 \cdot 2}$, namely, $b_{(y \cdot 2)(1 \cdot 2)}$. Therefore, the index error of $b_{y1 \cdot 2}$ should be identical with the index error of $b_{(y \cdot 2)(1 \cdot 2)}$. The latter can be obtained from the one-regressor case of Section 5.2; it is $\sqrt{1/m_{(1 \cdot 2)(1 \cdot 2)}}$. But by an argument parallel to that underlying (3.23), we see that $\sqrt{1/m_{(1 \cdot 2)(1 \cdot 2)}} = \sqrt{1/(m_{11} - m_{12}^2/m_{22})}$, which in turn may be written $\sqrt{m_{22}/(m_{11}m_{22} - m_{12}^2)}$.

Turning to the second slope, parallel arguments will show first that $\Delta_2 = \sqrt{1/m_{22}}$ is still feasible: accompanied by $\Delta_1 = 0$ it makes the parenthetical expression in (5.6) just equal to 1. The index error of $b_{y2 \cdot 1}$ is $\Delta_2^* = \sqrt{m_{11}/(m_{11}m_{22} - m_{12}^2)}$; accompanied by the compensating deviation $\Delta_{1(2*)} = (-m_{12}/m_{11})\Delta_2^*$, it also makes the parenthetical expression in (5.6) just equal to 1.

Noting that the denominator in the index error formulas, $m_{11}m_{22} - m_{12}^2$, equals $m_{11}m_{22}(1 - R_{21}^2)$, we may rewrite the index errors as

$$\Delta_1^* = \sqrt{1/[m_{11}(1 - R_{21}^2)]}, \qquad \Delta_2^* = \sqrt{1/[m_{22}(1 - R_{21}^2)]}. \quad (5.15)$$

In (5.15), we again see that if the sample variation of a regressor (m_{11} or m_{22}) is small, the imprecision which attaches to its slope will tend to be large. This is as it was in the one-regressor case and in the uncorrelated two-regressor case, but now the presence of auxiliary correlation introduces a new element. Correlation between the regressors tends to raise the imprecision of their slopes: As the auxiliary coefficient of determination R_{21}^2 approaches 1, the imprecision which attaches to the invidiual slopes grows large, *even if* m_{11} and m_{22} are large. One explanation is that the independent variation of the regressor is small. The conclusion that imprecision rises with R_{21}^2 is consistent with the intuitive argument that if the two regressors are highly correlated in the sample it should be difficult to reliably estimate their separate effects. We return to this point in Section 6.3.

Attention should also be given to the common sign of the compensating deviations $\Delta_{2(1*)} = (-m_{12}/m_{22})\Delta_1^*$ and $\Delta_{1(2*)} = (-m_{12}/m_{11})\Delta_2^*$. This sign depends only on the sign of the auxiliary correlation. If the regressors are positively correlated ($m_{12} > 0$), the compensating variation is negative: If we raise one slope we should lower the other slope to offset the increase in the sum of squared residuals ("the regressors are substitutes"). If the regressors are negatively correlated ($m_{12} < 0$), the compensating variation is positive: If we raise one slope we should raise the other to offset the increase in sum of squared residuals ("the regressors are complements").

Once again a geometric interpretation may be helpful. If we would plot $(\sum e^{*2} - \sum e^2)$—that is, $\Delta_1^2 m_{11} + \Delta_2^2 m_{22} + 2\Delta_1\Delta_2 m_{12}$—as a function of Δ_1 and Δ_2, we would still obtain an elliptic paraboloid—a bowl-shaped figure—opening upward, symmetric about the vertical axis, with vertex at the origin. But the orientation is different than in the uncorrelated case: Projecting onto the Δ_1, Δ_2 plane as in Figure 5.4, the family of homothetic ellipses, concentric about the origin, have axes which do not coincide with the coordinate axes.

Let us focus our attention on the index ellipse, that is, the contour for which $\left(\sum e^{*2} - \sum e^2\right) = 1$. We already know that it intersects the co-ordinate axes at the points $\pm(\sqrt{1/m_{11}}, 0)$ and $\pm(0, \sqrt{1/m_{22}})$, and also that it passes through the *compensated index error points* $\pm(\Delta_1^*, \Delta_{2(1*)})$ and $\pm(\Delta_{1(2*)}, \Delta_2^*)$. Further characterization of the ellipse is available from analytic geometry, in particular from material usually discussed under the heading "rotation of axes"—see, for example, Gay (1950, Chap. 16).

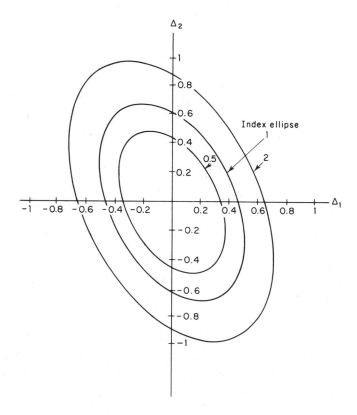

Figure 5.4   Contours of $\left(\sum e^{*2} - \sum e^2\right) = \Delta_1^2 m_{11} + \Delta_2^2 m_{22} + 2\Delta_1 \Delta_2 m_{12}.$

By considering a rotation of coordinate axes, it can be shown that an ellipse with equation $\Delta_1^2 m_{11} + \Delta_2^2 m_{22} + 2\Delta_1 \Delta_2 m_{12} = 1$ has the following features:

1. The ellipse is centered at the origin.

2. The major axis of the ellipse is the line through the origin with slope

$$\tan \theta = \frac{2m_{12}}{(m_{11} - m_{22}) - \sqrt{(m_{11} - m_{22})^2 + 4m_{12}^2}}, \qquad (5.16)$$

where $-90° \le \theta \le 90°$. It is readily confirmed that the denominator of (5.16) is always nonpositive, so that the sign of the slope depends only on the sign of the auxiliary correlation: When there is positive auxiliary correlation ($m_{12} > 0$), the slope is negative, so that the major axis runs in a general northwest-southeast direction (as in Figure 5.4); when there is negative auxiliary correlation ($m_{12} < 0$), the slope is positive, so that the major axis runs in a general southwest-northeast direction. It will be noted that the orientation of the index ellipse is in accord with the substitute-complement distinction made above.

3. The vertices of the index ellipse are at the points

$$\pm\sqrt{1/\lambda_1}\,(\cos \theta, \sin \theta), \qquad \pm\sqrt{1/\lambda_2}\,(-\sin \theta, \cos \theta) \qquad (5.17)$$

where

$$\lambda_1 = (\cos^2 \theta)m_{11} + 2(\cos \theta)(\sin \theta)m_{12} + (\sin^2 \theta)m_{22}, \qquad (5.18)$$

$$\lambda_2 = (\sin^2 \theta)m_{11} - 2(\cos \theta)(\sin \theta)m_{12} + (\cos^2 \theta)m_{22}. \qquad (5.19)$$

and $\lambda_1 \le \lambda_2$.

4. Thus the semilength of the major axis is $\sqrt{1/\lambda_1}$ and the semilength of the minor axis is $\sqrt{1/\lambda_2}$.

5. The area of the ellipse is $\pi(\sqrt{1/\lambda_1})(\sqrt{1/\lambda_2}) = \pi/\sqrt{m_{11}m_{22} - m_{12}^2}$, so that $1/\sqrt{m_{11}m_{22}(1 - R_{21}^2)} = 1/\sqrt{m_{11}m_{22} - m_{12}^2}$ may be taken as a measure of the joint imprecision which attaches to the least-squares slopes.

It is important to distinguish between the vertices and the compensated index error points. The vertices locate points on the ellipse with extreme (maximum or minimum) distance from the origin; the compensated index error points locate points on the ellipse with maximum abscissa or ordinate. These loci coincide if and only if $m_{12} = 0$, for then and only then do the principal axes of the ellipse coincide with the coordinate axes.

Example 5.2

The results of this section may be illustrated by the data of Example 2.5. First we consider an uncorrelated case, with $m_{11} = 32/6$ and $m_{22} = 17/6$ as in that example but with $m_{12} = 0$. These numbers were used to construct Figures 5.2 and 5.3. The index errors are

$$\Delta_1^* = \sqrt{1/m_{11}} = \sqrt{6/32} \approx 0.433,$$

$$\Delta_2^* = \sqrt{1/m_{22}} = \sqrt{6/17} \approx 0.594.$$

The index ellipse, namely, $\Delta_1^2(32/6) + \Delta_2^2(17/6) = 1$, is indicated in Figure 5.3 and is redrawn in Figure 5.5. Note that the principal axes coincide with the

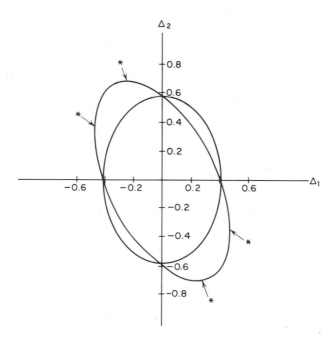

Figure 5.5   Index ellipses: Uncorrelated and correlated cases.
[* = Compensated index error points for correlated case.]

coordinate axes, so that the vertices coincide with the index error points $\pm(\sqrt{6/32}, 0)$ and $\pm(0, \sqrt{6/17})$. The area of the index ellipse is

$$\pi\sqrt{1/m_{11}}\sqrt{1/m_{22}} = \pi\sqrt{6/32}\sqrt{6/17} = 3\pi/2\sqrt{34} \approx 0.257\pi.$$

Example 5.3

We now turn to the correlated case of Example 2.5, where $m_{11} = 32/6$, $m_{22} = 17/6$, and $m_{12} = 10/6$. These data were used to construct Figure 5.4. The index errors are

$$\Delta_1^* = \sqrt{(17/6)/[(32/6)(17/6) - (10/6)^2]} = \sqrt{17/74} \approx 0.479,$$

$$\Delta_2^* = \sqrt{(32/6)/[(32/6)(17/6) - (10/6)^2]} = \sqrt{32/74} \approx 0.658.$$

The respective compensating deviations are

$$\Delta_{2(1\cdot)} = (-m_{12}/m_{22})\Delta_1^* = [-(10/6)/(17/6)]\sqrt{17/74}$$
$$= -(10/17)\sqrt{17/74} \approx -0.282,$$

$$\Delta_{1(2\cdot)} = (-m_{12}/m_{11})\Delta_2^* = [-(10/6)/(32/6)]\sqrt{32/74}$$
$$= -(5/16)\sqrt{32/74} \approx -0.206.$$

The compensated index error points are thus

$$\pm(\Delta_1^*, \Delta_{2(1\cdot)}) = \pm\frac{1}{\sqrt{74}}(\sqrt{17}, -10/\sqrt{17}) \approx \pm(0.479, -0.282),$$

$$\pm(\Delta_{1(2\cdot)}, \Delta_2^*) = \pm\frac{\sqrt{2}}{\sqrt{74}}(-5/4, 4) \qquad \approx \pm(-0.206, 0.658).$$

The index ellipse, namely, $\Delta_1^2(32/6) + \Delta_2^2(17/6) + 2\Delta_1\Delta_2(10/6) = 1$, is indicated in Figure 5.4 and redrawn in Figure 5.5. The compensated index error points are marked by asterisks. The following characteristics of the ellipse are computed; they may be confirmed by inspection of the figure.

$$\tan\theta = \frac{2(10/6)}{[(32/6) - (17/6)] - \sqrt{[(32/6) - (17/6)]^2 + 4(10/6)^2}}$$
$$= 20/(15 - 25) = -2$$

is the slope of the major axis. Thus $\theta = \arctan(-2) = -63°26'$ is the angle made by the major axis with respect to the horizontal axis, so that $\sin\theta = -2/\sqrt{5}$ and $\cos\theta = 1/\sqrt{5}$. Thus we compute

$$\lambda_1 = (1/5)(32/6) + (2)(1/\sqrt{5})(-2/\sqrt{5})(10/6) + (4/5)(17/6) = 60/30 = 2,$$

$$\lambda_2 = (4/5)(32/6) - (2)(1/\sqrt{5})(-2/\sqrt{5})(10/6) + (1/5)(17/6) = 185/30 = 37/6.$$

This implies that: The major axis has semilength $\sqrt{1/\lambda_1} = \sqrt{1/2} \approx 0.707$; its vertices are at $\pm\sqrt{1/\lambda_1}(\cos\theta, \sin\theta) = \pm\sqrt{1/2}(1/\sqrt{5}, -2/\sqrt{5}) \approx \pm(0.316, -0.632)$. The minor axis has semilength $\sqrt{1/\lambda_2} = \sqrt{6/37} \approx 0.403$; its vertices are at $\pm\sqrt{1/\lambda_2}(-\sin\theta, \cos\theta) = \pm\sqrt{6/37}(2/\sqrt{5}, 1/\sqrt{5}) \approx \pm(0.360, 0.180)$. The area of the ellipse is $\pi\sqrt{1/\lambda_1}\sqrt{1/\lambda_2} = \pi\sqrt{(1/2)(6/37)} = \pi\sqrt{3/37} = \pi/\sqrt{(32/6)(17/6)-(10/6)^2} = \pi/\sqrt{m_{11}m_{22} - m_{12}^2} \approx 0.285\pi$. In Figure 5.5 this index ellipse may be compared with that of the previous example to suggest the effect of auxiliary correlation. Note how the ellipse has been rotated, stretched, and expanded.

5.4.   INDEX ERRORS OF REGRESSION COEFFICIENTS: K-REGRESSOR CASE

We proceed to the K-regressor case discussed in Section 2.3, taking the variables to be measured as deviations about their sample means as in Section 3.3. The LSRF is $\hat{\mathbf{y}} = \mathbf{Xb}$, where $\mathbf{b} = (\mathbf{X'X})^{-1}\mathbf{X'y}$ is the $K \times 1$ least-squares slope vector. Suppose that instead of the LSRF we take the function $\mathbf{y^*} = \mathbf{X}(\mathbf{b} + \boldsymbol{\Delta})$, where $\boldsymbol{\Delta}$ is the $K \times 1$ vector of deviations of the pseudoslopes from the corresponding least-squares slopes. The residual vector from the pseudofunction is

$$\mathbf{e^*} = \mathbf{y} - \mathbf{y^*} = \mathbf{y} - (\mathbf{Xb} + \boldsymbol{\Delta}) = \mathbf{e} - \mathbf{X}\boldsymbol{\Delta}, \tag{5.20}$$

where $\mathbf{e} = \mathbf{y} - \mathbf{Xb}$ is the least-squares residual vector. Thus the sum of squared residuals from the pseudofunction will be

$$\mathbf{e^{*'}e^*} = (\mathbf{e} - \mathbf{X}\boldsymbol{\Delta})'(\mathbf{e} - \mathbf{X}\boldsymbol{\Delta}) = \mathbf{e'e} + \boldsymbol{\Delta'}\mathbf{X'X}\boldsymbol{\Delta}, \tag{5.21}$$

where $\mathbf{e'e}$ is the sum of squared residuals from the least-squares function, and we have used the fact that $\mathbf{X'e} = \mathbf{0}$ by the condition determining the least-squares function. The least-squares property of $\mathbf{b}$ is confirmed by (5.21): Since $\mathbf{X'X}$ is a positive definite matrix, $\mathbf{e^{*'}e^*}$ attains a minimum at $\boldsymbol{\Delta} = \mathbf{0}$.

Again we seek the index errors of the regression slopes, asking by how much we can change each slope from its least-squares value without raising the sum of squared residuals by more than 1. To find the index error of the first slope, say, we may maximize Δ_1^2, the square of the first element of $\boldsymbol{\Delta}$, subject to the constraint $\boldsymbol{\Delta'}\mathbf{X'X}\boldsymbol{\Delta} = 1$. It is convenient to define the $K \times 1$ vector $\boldsymbol{\alpha}$, which has a 1 as its first element and 0's elsewhere; then $\Delta_1 = \boldsymbol{\Delta'}\boldsymbol{\alpha} = \boldsymbol{\alpha'}\boldsymbol{\Delta}$ and $\Delta_1^2 = \boldsymbol{\Delta'}\boldsymbol{\alpha}\boldsymbol{\alpha'}\boldsymbol{\Delta} = \boldsymbol{\Delta'}\mathbf{A}\boldsymbol{\Delta}$, where $\mathbf{A} = \boldsymbol{\alpha}\boldsymbol{\alpha'}$ is a $K \times K$ matrix with 1 in the upper-left-hand corner and 0's elsewhere.

We form the Lagrangian expression $L = \Delta'A\Delta - \mu(\Delta'X'X\Delta - 1)$ and differentiate:

$$\partial L/\partial \Delta = 2A\Delta - 2\mu X'X\Delta, \tag{5.22}$$

$$\partial L/\partial \mu = -(\Delta'X'X\Delta - 1). \tag{5.23}$$

Setting these equal to zero gives

$$A\Delta - \mu X'X\Delta = 0, \tag{5.24}$$

$$\Delta'X'X\Delta = 1. \tag{5.25}$$

Multiplying (5.24) through by Δ' and solving gives $\mu = \Delta'A\Delta/\Delta'X'X\Delta$. which in view of (5.25) means that $\mu = \Delta'A\Delta = \Delta_1^2$. Using this value of μ and the fact that $A\Delta = \Delta_1\alpha$ because of the special structure of A and α, we may rewrite (5.24) as

$$\Delta_1(\alpha - \Delta_1 X'X\Delta) = 0. \tag{5.26}$$

For the nonzero Δ_1 of interest, the solution to (5.26) is given by setting the term in parentheses equal to zero, that is, by taking

$$\Delta_1\Delta = (X'X)^{-1}\alpha; \tag{5.27}$$

it can be shown that this locates a maximum. Now, in view of the special structure of α, the $K \times 1$ vector $(X'X)^{-1}\alpha$ is simply the first column of $(X'X)^{-1}$. Thus the first element of the maximizing $\Delta_1\Delta$, namely, Δ_1^2, is the first element of the first column of the inverse matrix. The index error of the first slope is just the square root of this.

Let us denote the i, j element of $(X'X)^{-1}$ as m^{ij}—this is consistent with our previous denotation of the i, j element of $X'X$ as m_{ij} (when variables are measured as deviations about the mean). Then our result for the K-regressor case is that the index error of the first slope is $\Delta_1^* = \sqrt{m^{11}}$. This is the maximum (absolute) deviation of the first slope from its least-squares value which is compatible with no more than a unit increase in the sum of squared residuals. Further, this deviation is to be accompanied by compensating variations in the remaining slopes, as defined by the remaining elements of the solution vector Δ in (5.27), namely, $m^{i1}/\sqrt{m^{11}} = (m^{i1}/m^{11})\sqrt{m^{11}}$ for $i = 2, \ldots, K$.

The argument provided for the first slope can, of course, be repeated for the other slopes. Thus we have, in fact, found that in the K-regressor case, the index error of the kth regression slope is given by

$$\Delta_k^* = \sqrt{m^{kk}}. \tag{5.28}$$

This is the maximum (absolute) deviation of the kth slope from its least-squares value compatible with no more than a unit increase in the sum of squared residuals; compatibility being assured by compensating variations in the other slopes, namely, $\Delta_{j(k*)} = (m^{jk}/m^{kk})\Delta_k^*$ for $j = 1, \ldots, k-1$, $k+1, \ldots, K$.

The reader will find it instructive to specialize these results to the cases $K = 1$ and $K = 2$ for comparison with our previous formulas. Interpretation of the index error in terms of sample variation of the regressor is given in Section 5.5, item 5, and in Section 6.3. A sketch of the geometric interpretation is given in Section 5.5, item 3.

5.5. SUPPLEMENTARY REMARKS

1. The approach taken in this chapter to measuring the precision of regression coefficients is hardly traditional. Nor are the concepts "index error" and "index ellipse" in common use. [The term "ellipsoide indicateur," which is mentioned by Malinvaud (1966, pp. 142–143), suggested this nomenclature.] Our approach has severe limitations. After all, the seriousness of a unit increase in the sum of squared residuals should depend on the level of that sum. In Section 6.5 we shall propose a modification of the index error which will eliminate that objection. But even so, the size of the sample should be relevant to measuring the precision of a regression coefficient. Ultimately, the limitation of the present approach is that it does not draw upon a complete stochastic model. Its saving grace, if any, lies in the relationship of our concepts to those of classical statistical inference. As we shall see in Chapter 7, the classical standard error of a regression coefficient can be factored into three components, one of which is the index error. Further, the index ellipse has the same shape as the boundary of the classical confidence region. We may say that the present approach focuses on one aspect of the classical treatment of precision to the neglect of the others. This may have some advantage as an analytical device; at the least, we will have provided a fresh interpretation of the conventional computations of standard errors, confidence regions, and test statistics.

It is also worth noting that plotting of the error sum of squares function, especially for nonlinear regression models, has recently become recommended practice—see, for example, Draper and Smith (1966, Chap. 10).

2. By working in terms of variables measured as deviations about their sample means we have in effect allowed the intercept of the pseudofunction to vary optimally in compensation for variations in the slopes. Working in terms of deviations about the means amounts to considering functions $y^* = (b_0 + \Delta_0) + \sum_{k=1}^{K}(b_k + \Delta_k)x_k$, where the $\Delta_1, \ldots, \Delta_K$ are arbitrary, but $(b_0 + \Delta_0) = \bar{y} - \sum_{k=1}^{K}(b_k + \Delta_k)\bar{x}_k$, that is, $\Delta_0 = -\sum_{k=1}^{K}\Delta_k \bar{x}_k$. Like the least-squares function, such pseudofunctions go through the point of sample means (and hence have zero sum of residuals). In effect, we have been considering rotations of the least-squares function about the sample mean point. While one could extend the analysis to cover translations as well, it is not hard to see that this would not change our conclusions: The maximum variation of a slope from its least-squares value consistent with no more than a unit increase in the residual sum of squares is accomplished by rotation without translation.

3. The geometric interpretation of the imprecision measures for the K-regressor case may be briefly suggested. The function $\Delta'X'X\Delta = 1$, where $X'X$ is a positive definite matrix, plots as an ellipsoid in K-dimensional space. This *index ellipsoid* is centered at the origin and its K principal axes run along K orthogonal eigenvectors of $X'X$. The semilength of each axis is given by the square root of the reciprocal of the corresponding eigenvalue; the major axis corresponds to the smallest eigenvalue. The orientation of the principal axes of the ellipsoid with respect to the coordinate axes may be studied in terms of a rotation accomplished by diagonalizing the symmetric matrix $X'X$ by an orthogonal matrix. The compensated index error points

$$\pm\sqrt{1/m^{11}}\begin{bmatrix} m^{11} \\ \vdots \\ m^{j1} \\ \vdots \\ m^{K1} \end{bmatrix}, \ldots, \pm\sqrt{1/m^{kk}}\begin{bmatrix} m^{1k} \\ \vdots \\ m^{jk} \\ \vdots \\ m^{Kk} \end{bmatrix}, \ldots, \pm\sqrt{1/m^{KK}}\begin{bmatrix} m^{1K} \\ \vdots \\ m^{jK} \\ \vdots \\ m^{KK} \end{bmatrix}$$

locate points on the ellipsoid which have maximum (absolute) values of the 1st, ..., kth, ..., Kth coordinates, respectively. The vertices locate points of extreme distance from the origin. The volume of the index ellipsoid is

$$\frac{(\sqrt{\pi})^K}{(K/2)\Gamma(K/2)}\sqrt{(1/\lambda_1)\cdots(1/\lambda_K)}=\frac{(\sqrt{\pi})^K}{(K/2)\Gamma(K/2)}\frac{1}{\sqrt{|X'X|}},$$

where the λ_k $(k = 1, \ldots, K)$ are the eigenvalues of $X'X$, $|X'X|$ is the determinant of $X'X$, and $\Gamma(K/2)$ is the gamma function

$$\Gamma(K/2) = \begin{cases} \left(\dfrac{K-2}{2}\right)! & \text{if } K \text{ is even,} \\[2ex] \dfrac{(K-2)(K-4)\cdots(1)\sqrt{\pi}}{2^{(K-1)/2}} & \text{if } K \text{ is odd.} \end{cases}$$

In the present approach, the index errors $\sqrt{m^{kk}} = \sqrt{1/m^{kk}}\,m^{kk}$ $(k = 1, \ldots, K)$ are proposed as measures of the imprecision which attaches to the individual regression slopes, and $1/\sqrt{|X'X|}$ as a measure of their joint imprecision.

The reader may find it instructive to specialize these results to the cases $K = 1$ and $K = 2$ for comparison with our earlier formulas. For a review of some of the mathematical tools used in the discussion of the K-regressor case, the following sources are suggested: Goldberger (1964, pp. 28–48) on eigenvalues, eigenvectors, definite matrices, and matrix differentiation; Scheffé (1959, pp. 371–411) on the geometric interpretation of matrix and vector algebra, including ellipsoids in particular; and Courant (1936, pp. 302–307, 637) on the volume of an ellipsoid.

4. The index ellipse describes the imprecision of the regression co-efficients. It is interesting to note a connection between its principal axes and the orthogonal auxiliary regressions which describe the sample co-variation among the regressors. We consider only the two-regressor case. The orthogonal regression of x_2 on x_1 has as its slope

$$b_p = \frac{2m_{12}}{(m_{11} - m_{22}) + \sqrt{(m_{11} - m_{22})^2 + 4m_{12}^2}} \tag{5.29}$$

—see (2.57). Comparing this with (5.16) we see that $b_p = -1/\tan\theta$, where $\tan\theta$ is the slope of the major axis of the index ellipse. The orthogonal regression runs perpendicular to the major axis of the index ellipse; that is, it runs along the minor axis. For a discussion of the relation between orthogonal regressions and principal axes, see Cramér (1951), pp. 275–276, 309–310). (*Caution:* Cramér is concerned not with the index ellipse but with one which is "inverse" to it.)

5. The following manipulations may clarify the role of auxiliary correlation in reducing precision in the general K-regressor case. Let the $K \times K$ regressor moment matrix be partitioned as

$$\mathbf{X'X} = \begin{bmatrix} m_{11} & m_{12} & \cdots & m_{1K} \\ m_{21} & m_{22} & \cdots & m_{2K} \\ \vdots & \vdots & & \vdots \\ m_{K1} & m_{K2} & \cdots & m_{KK} \end{bmatrix} = \mathbf{M} = \begin{bmatrix} m_{11} & \mathbf{M}_{12} \\ \mathbf{M}_{21} & \mathbf{M}_{22} \end{bmatrix} \quad (5.30)$$

where m_{11} is 1×1, $\mathbf{M}_{12}$ is $1 \times (K-1)$, $\mathbf{M}_{21} = \mathbf{M}'_{12}$, and $\mathbf{M}_{22}$ is $(K-1) \times (K-1)$. Then the first column of $(\mathbf{X'X})^{-1}$, the inverse of the regressor moment matrix, may be expressed in a conformably partitioned form as

$$\begin{bmatrix} m^{11} \\ m^{21} \\ \vdots \\ m^{K1} \end{bmatrix} = \begin{bmatrix} m^{11} \\ \mathbf{M}^{21} \end{bmatrix} = \begin{bmatrix} (m_{11} - \mathbf{M}_{12}\mathbf{M}_{22}^{-1}\mathbf{M}_{21})^{-1} \\ -\mathbf{M}_{22}^{-1}\mathbf{M}_{21}(m_{11} - \mathbf{M}_{12}\mathbf{M}_{22}^{-1}\mathbf{M}_{21})^{-1} \end{bmatrix}. \quad (5.31)$$

[To show this, one need only multiply (5.30) into (5.31): What results is $(1, 0, \ldots, 0)'$—the first column of the identity matrix. See also Goldberger (1964, p. 174). (*Caution:* In that reference the roles of 1 and 2 are the reverse of their present ones.)]

Now, consulting (3.47) we see that

$$m_{11} - \mathbf{M}_{12}\mathbf{M}_{22}^{-1}\mathbf{M}_{21} = m_{11}(1 - R^2_{1 \cdot 2, \ldots, K}), \quad (5.32)$$

where the long subscript indicates that the coefficient of determination refers to the auxiliary regression of x_1 on $x_2, \ldots, x_K$. Therefore, in view of (5.31) and (5.32), the squared index error of the first slope in the regression of y on $x_1, \ldots, x_K$ can be expressed

$$m^{11} = (m_{11} - \mathbf{M}_{12}\mathbf{M}_{22}^{-1}\mathbf{M}_{21})^{-1} = 1/[m_{11}(1 - R^2_{1 \cdot 2, \ldots, K})]. \quad (5.33)$$

The role played by x_1 in this development can be reassigned to any x_k: We have, in fact, established that in the K-regressor case, the index error of each b_k ($k = 1, \ldots, K$) may be expressed

$$\Delta_k^* = \sqrt{m^{kk}} = \sqrt{1/[m_{kk}(1 - R_{k \cdot 1,\ldots,k-1,k+1,\ldots,K}^2)]}, \qquad (5.34)$$

where the long subscript indicates that the coefficient of determination refers to the auxiliary regression of x_k on all the other regressors. When this coefficient of determination is high, the independent variation of x_k will tend to be small, and the index error of b_k will tend to be large. For further discussion of this point, see Section 6.3.

For future reference, another result will also be useful. By the cofactor method of computing elements in an inverse matrix, $m^{11} = |\mathbf{M}_{22}|/|\mathbf{M}|$. Taking this together with (5.33) we obtain the following expression for the determinant of the regressor moment matrix:

$$|\mathbf{X'X}| = |\mathbf{M}| = |\mathbf{M}_{22}|/m^{11} = m_{11}|\mathbf{M}_{22}|(1 - R_{1 \cdot 2,\ldots,K}^2). \qquad (5.35)$$

More generally, we may write for any k,

$$|\mathbf{X'X}| = |\mathbf{M}| = m_{kk}|\mathbf{M}_{k'k'}|(1 - R_{k \cdot 1,\ldots,k-1,k+1,\ldots,K}^2), \qquad (5.36)$$

where $\mathbf{M}_{k'k'}$ denotes the $(K - 1) \times (K - 1)$ matrix which remains when the kth row and column of $\mathbf{M}$ are deleted.

Chapter 6

Precision of Estimation: Further Results

6.1. INTRODUCTION

IN THIS CHAPTER WE APPLY THE CONCEPTS DEVELOPED in Chapter 5 to a variety of problems which arise in the application of regression analysis to empirical data. The topics considered include precision of estimation of combinations of regression coefficients, the problem of multicollinearity, and forecasting.

6.2. LINEAR COMBINATIONS OF REGRESSION COEFFICIENTS

In some empirical applications our major interest is not in the individual regression coefficients but in combinations of them. For example, for some questions of fiscal policy it is the *differences* between the marginal propensities to consume at various income levels which are relevant rather than the marginal propensities themselves. Thus assuming a linear PRF, say, $E(C \mid H, L) = \beta_0 + \beta_H H + \beta_L L$ (where C = consumption, H = income of high earners, and L = income of low earners), we would be concerned with the difference, $\beta_L - \beta_H$, rather than with β_L and β_H separately. Clearly the difference between the least-squares slopes, namely, $b_L - b_H$, estimates that difference, but it would be useful to have some idea of the reliability which attaches to this estimate. Another example is provided in the context of development policy, where the question of returns to scale in production is important. Here it is the sum of the input elasticities rather than the individual elasticities which is relevant; in the logarithmic version of the Cobb-Douglas production function this is given by the *sum* of regression coefficients. The sum of the least-squares slopes will estimate that sum; again it would be useful to have some idea of the reliability which attaches to the estimated sum.

Now, differences and sums are just special cases of *linear combinations.* That is, $\gamma = \sum_{k=0}^{K} a_k \beta_k$, *where the a_k's are given numbers*, is a linear combination of the population regression coefficients; $c = \sum_{k=0}^{K} a_k b_k$ is the corresponding linear combination of the sample regression coefficients and hence the natural estimate of γ. For example, the γ defined by $a_1 = 1$, $a_2 = -1$, all other a_k's $= 0$, is the difference between the first and second regression slopes; the γ defined by $a_0 = 0$, $a_1 = \cdots a_K = 1$, is the sum of the regression slopes. We establish results on precision for the general case of a linear combination of regression coefficients; these can be specialized as needed in practice.

At first glance it might seem that the imprecision of a linear combination should be a linear combination of the imprecisions of the individual elements. But life is not that simple. From other contexts it is clear that we may be quite uncertain about the value of each of two parameters and yet be quite certain about the difference between them. For an illustration, it is said that even if the estimates of annual GNP in the national income accounts contain substantial error, the year-to-year *changes* in GNP may still be quite accurately estimated. This assertion rests on the belief that there are sources of error which are positively correlated over time, that is, more or less repeat themselves each year. (Underreporting on tax returns is one such source.) This illustration is in fact rather suggestive: As we shall see, the reliability of c as compared with that of the b's also depends upon correlation.

With this as background, we proceed to measure the imprecision of linear combinations of regression coefficients. The index error of the linear combination $c = \sum_{k=0}^{K} a_k b_k$ is, naturally, defined as the maximum (absolute) amount by which it can be changed from its least-squares value without raising the sum of squared residuals by more than 1. Taking the K-regressor case, we add to the notation of Section 5.4 the following: Let $\mathbf{a}' = (a_1, \ldots, a_K)$ be the $1 \times K$ vector of coefficients of the linear combination and $\boldsymbol{\beta}$ be the population slope vector; then $\gamma = \mathbf{a}'\boldsymbol{\beta}$ is the linear combination whose value concerns us.

The least-squares function estimates the value of this linear combination as $c = \mathbf{a}'\mathbf{b}$, where $\mathbf{b}$ is the least-squares slope vector. If we take a pseudo-function $\mathbf{y}^* = \mathbf{X}(\mathbf{b} + \boldsymbol{\Delta})$, then the estimated linear combination becomes

$\mathbf{a}'(\mathbf{b} + \boldsymbol{\Delta})$, which deviates from the least-squares value by $\mathbf{a}'\boldsymbol{\Delta} = \delta$, say. (Note that many different $\boldsymbol{\Delta}$'s can yield the same δ.) The increase in the sum of squared residuals attributable to taking the pseudofunction rather than the least-squares function is again $\boldsymbol{\Delta}'\mathbf{X}'\mathbf{X}\boldsymbol{\Delta}$. For the index error of an individual b_k we maximized Δ_k^2 subject to $\boldsymbol{\Delta}'\mathbf{X}'\mathbf{X}\boldsymbol{\Delta} = 1$; but now we maximize $\delta^2 = (\mathbf{a}'\boldsymbol{\Delta})^2 = \boldsymbol{\Delta}'\mathbf{a}\mathbf{a}'\boldsymbol{\Delta}$ subject to the same constraint.

Thus we form the Lagrangian expression $L = \boldsymbol{\Delta}'\mathbf{a}\mathbf{a}'\boldsymbol{\Delta} - \mu(\boldsymbol{\Delta}'\mathbf{X}'\mathbf{X}\boldsymbol{\Delta} - 1)$ and differentiate:

$$\partial L/\partial \boldsymbol{\Delta} = 2\mathbf{a}\mathbf{a}'\boldsymbol{\Delta} - 2\mu\mathbf{X}'\mathbf{X}\boldsymbol{\Delta} = 2(\mathbf{a}\mathbf{a}' - \mu\mathbf{X}'\mathbf{X})\boldsymbol{\Delta}, \tag{6.1}$$

$$\partial L/\partial \mu = -(\boldsymbol{\Delta}'\mathbf{X}'\mathbf{X}\boldsymbol{\Delta} - 1). \tag{6.2}$$

Setting (6.1) at zero and premultiplying through by $\boldsymbol{\Delta}'$ we find

$$\boldsymbol{\Delta}'\mathbf{a}\mathbf{a}'\boldsymbol{\Delta} = \mu\boldsymbol{\Delta}'\mathbf{X}'\mathbf{X}\boldsymbol{\Delta}. \tag{6.3}$$

Setting (6.2) at zero and inserting in (6.3) we find

$$\boldsymbol{\Delta}'\mathbf{a}\mathbf{a}'\boldsymbol{\Delta} = \mu. \tag{6.4}$$

Again, setting (6.1) at zero but now premultiplying through by $\mathbf{a}'(\mathbf{X}'\mathbf{X})^{-1}$ we find $[\mathbf{a}'(\mathbf{X}'\mathbf{X})^{-1}\mathbf{a}\mathbf{a}' - \mu\mathbf{a}'(\mathbf{X}'\mathbf{X})^{-1}(\mathbf{X}'\mathbf{X})]\boldsymbol{\Delta} = 0$ or $[\mathbf{a}'(\mathbf{X}'\mathbf{X})^{-1}\mathbf{a} - \mu](\mathbf{a}'\boldsymbol{\Delta}) = 0$, which for the nonzero $\mathbf{a}'\boldsymbol{\Delta}$ of interest implies

$$\mathbf{a}'(\mathbf{X}'\mathbf{X})^{-1}\mathbf{a} = \mu. \tag{6.5}$$

Combining (6.4) and (6.5) we find

$$\boldsymbol{\Delta}'\mathbf{a}\mathbf{a}'\boldsymbol{\Delta} = \mathbf{a}'(\mathbf{X}'\mathbf{X})^{-1}\mathbf{a} \tag{6.6}$$

for the solution. It can be shown that this locates a maximum. Thus we have found that the maximum (absolute) deviation of the linear combination from its least-squares value consistent with the constraint is

$$\delta^* = \sqrt{\mathbf{a}'(\mathbf{X}'\mathbf{X})^{-1}\mathbf{a}} = \sqrt{\sum_{j=1}^{K}\sum_{k=1}^{K} a_j a_k m^{jk}}, \tag{6.7}$$

which is thus the index error of $c = \mathbf{a}'\mathbf{b}$.

To obtain a feel for this result consider the case in which $a_i = 1$ while all other elements of $\mathbf{a}$ are 0. According to (6.7) the index error of the resulting linear combination is simply $\sqrt{m^{ii}}$. But this is as it should be: If all a_k's are 0 except for a_i, which is 1, then the "linear combination of regression coefficients $c = \sum_{k=0}^{K} a_k b_k$" is in fact just the single regression coefficient b_i; that $\sqrt{m^{ii}}$ is the index error of b_i we know from Chapter 5.

To proceed, we confine our attention until further notice to the two-regressor case, again treating variables as deviations about their sample means. The linear combinations of regression coefficients being considered are now $a_1 b_1 + a_2 b_2$, and the index error formula (6.7) may be written explicitly as

$$\sqrt{(a_1^2 m_{22} + a_2^2 m_{11} - 2a_1 a_2 m_{12})/(m_{11} m_{22} - m_{12}^2)}. \tag{6.8}$$

In particular, taking $a_1 = a_2 = 1$ we find that the index error of the sum of the two slopes is

$$\sqrt{(m_{22} + m_{11} - 2m_{12})/(m_{11} m_{22} - m_{12}^2)}; \tag{6.9}$$

while taking $a_1 = 1$ and $a_2 = -1$ we find that the index error of the difference of the two slopes is

$$\sqrt{(m_{22} + m_{11} + 2m_{12})/(m_{11} m_{22} - m_{12}^2)}. \tag{6.10}$$

Now, when the regressors are positively correlated, $m_{12} > 0$, so that the index error of the sum tends to be small while the index error of the difference tends to be large. This is in accord with common sense: When two variables move positively together in a sample, it should be easy to estimate their combined effect upon the conditional expectation of y, but hard to isolate their separate effects, and even harder to estimate the difference between their separate effects. On the other hand, when the regressors are negatively correlated, $m_{12} < 0$, and the index error of the sum tends to be large while the index error of the difference tends to be small. In the special case where the regressors are uncorrelated, $m_{12} = 0$, and both (6.9) and (6.10) reduce to $\sqrt{(m_{22} + m_{11})/(m_{11} m_{22})} = \sqrt{(1/m_{11}) + (1/m_{22})}$, which is just the square root of the sum of squared index errors of the individual regression coefficients.

Having considered the sum and the difference let us return to more general linear combinations of b_1 and b_2. We see from (6.8) that given the sample data (the m_{jk}'s) the index error of a linear combination depends upon the coefficients of the linear combination (the a_k's). Thus for a given sample there may be some linear combinations which can be estimated quite precisely and other linear combinations which can be estimated with very little precision. A critical factor in determining the imprecision of a linear combination is the *ratio* of its coefficients.

To concentrate on this factor we proceed as follows. Consider all linear combinations whose coefficients satisfy the normalization

$$a_1^2 + a_2^2 = h, \tag{6.11}$$

where h may be any positive constant. Of all such linear combinations, which has the smallest index error? We form the Lagrangian

$$L = a_1^2 m_{22} + a_2^2 m_{11} - 2a_1 a_2 m_{12} - \mu(a_1^2 + a_2^2 - h)$$

and differentiate with respect to a_1 and a_2. Setting these derivatives equal to 0 and then eliminating μ yields the following quadratic equation in the ratio a_1/a_2 :

$$m_{12}(a_1/a_2)^2 - (m_{11} - m_{22})(a_1/a_2) - m_{12} = 0. \tag{6.12}$$

Of the two roots to this equation, the one which minimizes

$$a_1^2 m_{22} + a_2^2 m_{11} - 2a_1 a_2 m_{12}$$

subject to (6.11) is

$$a_1/a_2 = [(m_{11} - m_{22}) + \sqrt{(m_{11} - m_{22})^2 + 4m_{12}^2}]/2m_{12}. \tag{6.13}$$

We conclude that the linear combinations whose coefficients stand in the ratio (6.13) are the ones whose index errors tend to be small.

Can we obtain a geometric interpretation of this result? Let $c = a_1 b_1 + a_2 b_2$ be the value of the linear combination of the least-squares slopes; the value of the linear combination of the pseudoslopes is then $a_1(b_1 + \Delta_1) + a_2(b_2 + \Delta_2) = c + \delta$, where $\delta = a_1 \Delta_1 + a_2 \Delta_2$. In the Δ_1, Δ_2 plane the function $\delta = a_1 \Delta_1 + a_2 \Delta_2$, which gives the deviation from the least-squares value of the linear combination, plots as a straight line, namely, $\Delta_2 = -(a_1/a_2)\Delta_1 + (\delta/a_2)$. In particular, the straight line through the origin with slope $-a_1/a_2$ locates all the pairs Δ_1, Δ_2 which make the deviation of the value of the linear combination from its least-squares value zero; that is, it locates all the pseudofunctions for which the linear combination attains its least-squares value. As we consider various values of δ, that is, deviations of the value of the linear combination from its least-squares value, we are effectively considering parallel shifts in this line. The question "How much can the value of the linear combination be varied from its least-squares value without raising the sum of squared residuals by more than 1?" is thus equivalent to the question "How far can we shift the line representing this linear combination in a parallel manner without going completely outside the index ellipse?"

Now, in the Δ_1, Δ_2 plane, the linear combinations whose coefficients stand in the optimal ratio of (6.13) are depicted by the straight lines with slope

$$-a_1/a_2 = -[(m_{11} - m_{22}) + \sqrt{(m_{11} - m_{22})^2 + 4m_{12}^2}]/2m_{12}$$

$$= 2m_{12}/[(m_{11} - m_{22}) - \sqrt{(m_{11} - m_{22})^2 + 4m_{12}^2}]. \quad (6.14)$$

Comparing this slope with (5.16), we see that it is just $\tan \theta$, the slope of the major axis of the index ellipse. This indicates that the linear combinations which are most precisely estimable are those which run parallel to the major axis of the index ellipse. The geometric interpretation of this may be seen in Figure 5.5. Consider the index ellipse of the correlated case, say: Clearly parallel shifts of the major axis will soon run entirely outside the index ellipse. On the other hand, parallel shifts of the minor axis can go on for quite a while before running outside the index ellipse. Broadly speaking, we may say that the more parallel to the major axis a linear combination is, the more precisely it is estimated.

Some simple applications may clarify this result. The *sum* linear combination ($a_1 = a_2 = 1$) has $-a_1/a_2 = -1$ and hence plots as a line running from northwest to southeast (at a $-45°$ angle); the *difference* linear combination ($a_1 = -a_2 = 1$) has $-a_1/a_2 = 1$ and hence plots as a line running from southwest to northeast (at a $45°$ angle). Suppose that the auxiliary correlation is positive ($m_{12} > 0$); as we have seen in Section 5.3, this means that the major axis will be negatively sloped, running broadly northwest to southeast. The sum combination is more parallel to this than is the difference combination and hence should have a smaller index error, and this is quite what we found in (6.8) to (6.10). On the other hand, suppose that the auxiliary correlation is negative ($m_{12} < 0$); as we have seen in Section 5.3, this means that the major axis will be positively sloped, running broadly southwest to northeast. The sum linear combination is less parallel to this than is the difference linear combination and hence should have a larger index error; again this corresponds to our finding from (6.8) to (6.10).

Indeed, the present approach may even "explain" why the index error of the first slope in Example 5.3 was smaller than that of the second slope. The first slope can be represented as the linear combination with $a_1 = 1$, $a_2 = 0$, which plots as a line with slope $-a_1/a_2 = -\infty$, that is, as a vertical line; while the second slope can be expressed as the linear combination with $a_1 = 0$, $a_2 = 1$, which plots as a line with slope $-a_1/a_2 = 0$, that is, as a horizontal line. The major axis of the index ellipse ran at an angle of $-63°26'$, and this is indeed closer to the vertical than to the horizontal.

Another application of the present approach is to the problem of *forecasting*. By a forecast with the LSRF we simply mean the value of the function for some specified set of values of the regressors: The forecast of y for $x_1 = x_{1*}, \ldots, x_K = x_{K*}$ is simply the calculated value $\hat{y}_* = \sum_{k=1}^{K} b_k x_{k*}$ (variables still being measured as deviations about their sample means). This is a plausible method of forecasting. After all, the natural predictor of a random variable given the values of other variables is its conditional expectation given those variables and our LSRF is an estimate of the conditional expectation.

Recognizing that our forecast is simply a linear combination of regression coefficients, namely, the one with $a_1 = x_{1*}, \ldots, a_K = x_{K*}$, we may ask: "For which sets of values of the x's can the most reliable forecasts be made?" It is easy to resolve this question for the present two-regressor case. According to (6.13) and (6.14) index errors will tend to be small for forecasts where $-x_{1*}/x_{2*} = \tan \theta$, that is, for $x_{2*} = (-1/\tan \theta)x_{1*}$. Now, in Section 5.5, item 4, we found that $(-1/\tan \theta)$ was the slope of the orthogonal auxiliary regression of x_2 on x_1. Thus our present result means that forecasts are most precise at values of the regressors which lie along the orthogonal auxiliary regression line. Since the sample values of the regressors tended to lie along that line, our present result is in accord with the commonsense notion that forecasting is most reliable when it falls within the range of experience. By the range of experience is here meant the regressor relations which prevailed in the sample.

6.3. MULTICOLLINEARITY

At various points we have seen that auxiliary correlation—correlation among the regressors in the sample—can hinder regression analysis by reducing the precision of individual regression coefficients. At the extreme when there is some *perfect* auxiliary correlation, it is quite impossible to estimate the individual regression coefficients. To see this for the two-regressor case, simply note that (3.8) and (3.9) can be solved uniquely if and only if $m_{11}m_{22}^2 - m_{12}^2 \neq 0$, that is, if and only if $m_{11}m_{22}(1 - R_{21}^2) \neq 0$. Perfect correlation between x_1 and x_2 means that $R_{21}^2 = 1$, which violates the condition. For the K-regressor case, recognize first that to solve the normal equations (2.35) by any method—for example, Cramer's rule or matrix inversion—we have effectively to divide by the determinant of the moments on the left side of (2.35). As was shown in (5.36), this determinant is proportional to $(1 - R_{k \cdot 1,\ldots,k-1,k+1,\ldots,K}^2)$ for any k, where the long subscript identifies the coefficient of determination in the auxiliary regression of x_k on all the other regressors. If any auxiliary correlation is perfect, the determinant will be zero and the normal equations will have no unique solution. (It is interesting to note that this can happen without any *pair* of regressors being perfectly correlated.) Alternatively, in terms of the discussion of Section 2.4, item 3, if there exists a nontrivial linear combination of the regressor vectors which equals zero, then one regressor vector can be expressed as an exact linear function of the others. The auxiliary regression of that regressor on the others will then provide a perfect fit; that is, the coefficient of determination will be 1.

We shall call this situation, where there is some perfect auxiliary correlation so that some regressor is in the sample an exact linear function of some others, *exact multicollinearity*. It does not occur often in practice, and of course when it does occur we should soon become aware of it as we attempt to solve the normal equations. However, situations very close to it do occur very often in practice.

Following Johnston (1963, p. 201), we define *multicollinearity* as the situation "which arises when some or all of the explanatory variables are so highly correlated one with another that it becomes very difficult, if not impossible, to disentangle their influences and obtain a reasonably precise estimate of their [separate] effects." It is useful to recognize the following points. Multicollinearity is a matter of degree rather than of all or nothing; we say that it is present when some auxiliary coefficients of determination are "high." Multicollinearity is a property of the sample data and not of the population; thus no clear meaning attaches to the phrase "testing for multicollinearity." Exact multicollinearity is a special case of multicollinearity: the one to which the "impossible" phrase of Johnston's statement refers.

While in economic data exact multicollinearity may be rare, multicollinearity is very common. High auxiliary correlations are frequent. Indeed, the rather cynical comment has been made that at the same time one researcher is regressing some y on some $x_1, \ldots, x_K$ and hoping to find a high R^2 in this regression, somewhere else in the world another researcher is

regressing one of those x's on the others and hoping to find a high R^2 in
that regression. Both may well be successful.

The general interdependence of economic phenomena may easily result
in the appearance of approximate linear relationships among regressors.
This possibility is most apparent in time series, because so many economic
time series do move closely together over the business cycle or over a sec-
ular growth path. It may occur in cross sections as well; households with
high incomes are likely to have high liquid assets, large amounts of
durables, and so forth.

The implications of multicollinearity for precision of individual regression coefficients are implicit in much of the discussion of the preceding sections. Heuristically it is clear why multicollinearity makes it
difficult to obtain reasonably precise estimates of the separate effects of
regressors. Consider the problem of trying to estimate the separate effects
of income and assets upon consumption. If we were in a laboratory situation
it would be feasible to conduct a controlled experiment. We would study
families which all had the same income level and measure how consumption
varies with their liquid asset holdings, then take another group of families
which all had a different common income level and measure how consumption varies with their liquid asset holdings, and so on. The income
effect would be manifested in shifts of the consumption-asset relation from
subsample to subsample. In this controlled situation there would be no
correlation between assets and income within each subsample, nor, if
average assets were kept the same in all the subgroups, would there be any
correlation between assets and income over the full sample. [Recall that
mean independence implies uncorrelatedness, as was shown in (2.4).]
But it is precisely this kind of controlled experimental design which is
lacking in real-world samples.

A quantitative formulation is desirable and is now readily available.
We have seen in (5.34) that the index error of a regression slope b_k is
expressible as

$$\Delta_k^* = \sqrt{1/[m_{kk}(1 - R^2_{k \cdot 1, \ldots, k-1, k+1, \ldots, K})]}. \qquad (6.15)$$

Examining this expression, we see that it tends to be large—so that the
imprecision which attaches to b_k will tend to be large—when the sample
variation of the corresponding regressor is small and/or the auxiliary
correlation (measured by the R^2 in the denominator) of that regressor on
all the others is large. This constitutes a straightforward generalization of
the result we obtained for the two-regressor case in (5.15).

Further, we have noted in Section 5.5, item 3, that $1/\sqrt{|\mathbf{X}'\mathbf{X}|}$ is a measure of the volume of the index ellipsoid. Now as we have seen in (5.35), $|\mathbf{X}'\mathbf{X}|$ will tend to be small if some auxiliary correlation is large. Thus the volume of the index ellipsoid will tend to be large when there is large auxiliary correlation. In that situation, widely different Δ's will fall within the index ellipsoid, which means that widely different sets of slopes will be "consistent" with the data, which in turn means that we shall be very uncertain about their respective values. This suggests again how multicollinearity tends to raise imprecision; it is a straightforward generalization of the result for the two-regressor case which was illustrated in Example 5.3.

Still another view may be illustrated in the two-regressor case. Suppose that the least-squares regression function is $\hat{y} = b_{y1 \cdot 2} x_1 + b_{y2 \cdot 1} x_2$, while in the sample, $x_{t2} = p x_{t1}$ approximately for every t. Then the pseudo-function $y_t^* = (b_{y1 \cdot 2} + \Delta_1) x_{t1} + (b_{y2 \cdot 1} + \Delta_2) x_{t2}$, where Δ_2 is any number and $\Delta_1 = -p\Delta_2$, will yield $y_t^* = \hat{y}_t$ approximately for every t in the sample, and hence will fit almost as well as the least-squares function.

When we are faced with severe multicollinearity, we may say that the sample is simply not rich enough in independent variation of the regressors to provide us with reliable estimates of their individual effects. To be sure, it may not prevent us from obtaining reasonably precise estimates of certain combinations of their effects. The analysis of Section 6.2 is relevant in this respect. We may then succeed in the face of multicollinearity if we ask less of the sample, that is, lower our aspiration level from seeking the individual regression coefficients to seeking only certain functions of them.

Alternatively we may succeed if we provide further information, asking not what the sample can do for us but what we can do for the sample. The standard approach to the problem of multicollinearity, indeed, is to utilize additional information as an aid in estimation. This additional information can consist of knowledge of the values of some regression coefficients, ratios among them, and so forth. It may stem from economic theory, or from other samples; Goldberger (1964, pp. 255–262) discusses various methods for incorporating such "extraneous" information. It turns out that even uncertain information—and even incorrect information—can reduce the imprecision of estimation. To give a taste of the procedure, we cite a simple example, drawn from Klein and Goldberger (1955, pp. 57–62). Attempts to estimate the consumption function $C = \beta_0 + \beta_1 W + \beta_2 P + \beta_3 A$ (where C = consumption, W = wage income, P = nonwage–nonfarm income, and A = farm income) from U.S. time-series data were futile because the three income variables moved so closely together. However, on the basis of some cross-section studies, it was estimated that

$\beta_2 = 0.75\beta_1$ and $\beta_3 = 0.625\beta_1$. The consumption function was rewritten $C = \beta_0 + \beta_1 W + 0.75\beta_1 P + 0.625\beta_1 A = \beta_0 + \beta_1 Y^*$, where Y^* is the weighted income variable $W + 0.75P + 0.625A$. Then the simple regression of C on Y^* in the time series estimated β_1, which, supplemented by the cross-section information, then estimated β_2 and β_3.

6.4. PRECONCEPTIONS OF THE POPULATION REGRESSION FUNCTION

It was suggested in Section 5.1 that our analysis of precision might help in judging whether a preconception about the population regression coefficients was compatible with the estimated sample regression coefficients. We may now take up this suggestion, considering preconceptions of a particular type. In some applications it is of interest to test the preconception that a population regression slope is zero, that is, that the conditional expectation of the regressand does not vary with the corresponding regressor, that is, "that the regressor does not belong in the PRF." Indeed, it is often of interest to test the joint preconception that *all* the regression slopes are zero, that is, that the conditional expectation of the regressand does not vary with any of the regressors, that is, that the PRF is constant. While it may seem paradoxical, it is true that there are circumstances in which it is reasonable to accept each of the preconceptions $\beta_1 = 0, \ldots, \beta_K = 0$ taken separately, but unreasonable to accept the joint preconception $\beta_1 = \cdots = \beta_K = 0$.

Even without a formal statistical model, the resolution of this paradox can be suggested by use of our index ellipse approach. Consider the two-regressor case of Section 5.3. Let us agree to say that a preconception is consistent with the data if it falls in or on the index ellipse, inconsistent otherwise. (The essence of the argument would not be changed if we insisted on another ellipse, $\sum e^{*2} - \sum e^2 = $ constant, rather than on the index ellipse $\sum e^{*2} - \sum e^2 = 1$.) That is, the preconception $\beta_1 = 0$ would be consistent with the data if there were *some* point in or on the index ellipse which had abscissa $\Delta_1 = -b_{y1 \cdot 2}$; the preconception $\beta_2 = 0$ would be consistent with the data if there were *some* point in or on the index ellipse which had ordinate $\Delta_2 = -b_{y2 \cdot 1}$. But the joint preconception $\beta_1 = 0$ *and* $\beta_2 = 0$ would be consistent with the data if *the* point with abscissa $\Delta_1 = -b_{y1 \cdot 2}$ and ordinate $\Delta_2 = -b_{y2 \cdot 1}$ fell in or on the index ellipse. It is not hard to see that the first two conditions can hold without the third holding.

In that event, the situation might be described as follows: It is possible that the conditional expectation of y does not vary with the first regressor—

but then it varies with the second. Alternatively, it is possible that the conditional expectation of y does not vary with the second regressor—but then it varies with the first. It is not possible that it varies with neither the first nor the second regressor.

We have discussed preconceptions which specify that certain population regression coefficients are zero, because they are frequently met. The spirit of this section will apply to the more general case where the preconception specifies $\beta_k = \beta_k^*$ $(k = 1, \ldots, H; 1 \leq H \leq K)$.

6.5. RADEX ERRORS OF REGRESSION COEFFICIENTS

There is something disturbing about the index approach as developed so far. Implicit in our results is the fact that the index error of the coefficient of a regressor necessarily rises as other (correlated) regressors are introduced into the LSRF. For example, the index error of the coefficient of x_1 in the regression of y on x_1 and x_2, $\sqrt{1/[m_{11}(1 - R_{21}^2)]}$, must exceed the index error of that coefficient in the regression of y on x_1 alone, $\sqrt{1/m_{11}}$, the only proviso being $R_{2i}^2 \neq 0$. This implication runs counter to commonsense notions of precision. Why, after all, should the imprecision which attaches to a regression slope inevitably increase when another regressor is added?

Common sense is correct here. It must be admitted that the index error is inappropriate for comparing the imprecision of a certain regression coefficient obtained in different functions fit to the same body of data. (Nor have we used it for this purpose; Examples 5.2 and 5.3 referred to the same function fitted to different bodies of data.) The source of the difficulty is not hard to find. The index error tells us what it takes to raise the sum of squared residuals from its minimum by just 1. Now, the seriousness with which we view a unit increase in the sum of squared residuals should depend upon the level of $\sum e^2$ itself. Presumably, a unit increase is less serious when the level is high. When we add another regressor we generally reduce $\sum e^2$ (at worst, it may remain unchanged). Some account should be taken of this shift in level. A reduction in $\sum e^2$ should tend to bring about an overall increase in precision. To put it in a slightly different way, the index error takes account only of the sample variation of the regressors (and their covariation), while an appropriate measure of precision should take some account of the variation of the regressand (and its covariation with the regressors).

To meet this objection, we now propose the *radex error* as a measure of the imprecision of a regression coefficient. The radex error π_k^* of a regression coefficient b_k is defined as the maximum (absolute) amount by which we can deviate from its least-squares value without raising the sum of squared residuals by more than 100 *per cent* of its least-squares value, that is, without making $\sum e^{*2} - \sum e^2 > \sum e^2$. Examination of the definition of the radex error in connection with (5.2), (5.6), or (5.21) shows immediately that the radex error is simply $\sqrt{\sum e^2}$ times the index error:

$$\pi_k^* = \sqrt{\sum e^2} \Delta_k^*. \tag{6.16}$$

In terms of the radex error an analysis of the effect of adding regressors upon precision would run as follows. If we take the regression of y on x_1 alone, we get $\sqrt{\sum e_y^2 \cdot 1 / m_{11}}$ for the radex error of the x_1 slope, while if we take the regression of y on x_1 and x_2 together, we get $\sqrt{\sum e_y^2 \cdot 12 / [m_{11}(1 - R_{21}^2)]}$ for the radex error of the x_1 slope. (We are resorting to the explicit notation of Section 4.2: $\sum e_y^2 \cdot 1$ is the "$\sum e_2$" of the simple regression and $\sum e_y^2 \cdot 12$ is the "$\sum e^2$" of the multiple regression.) A comparison of the two radex errors shows that the second will be smaller—so that adding a regressor reduces imprecision—if and only if $(\sum e_y^2 \cdot 12 / \sum e_y^2 \cdot 1) < (1 - R_{21}^2)$. This condition cuts in the right direction: It says that a sufficiently large reduction in the residual sum of squares can compensate for auxiliary correlation. Continuing, we recognize that $\sum e_y^2 \cdot 12 = (1 - R_y^2 \cdot 12) \sum y^2$ and that $\sum e_y^2 \cdot 1 = (1 - R_y^2 \cdot 1) \sum y^2$, whence the condition for reduction of imprecision can be restated as $(1 - R_y^2 \cdot 12)/(1 - R_y^2 \cdot 1) < (1 - R_{21}^2)$. But $(1 - R_y^2 \cdot 12)/(1 - R_y^2 \cdot 1) = 1 - R_{(y \cdot 1)(2 \cdot 1)}^2$, where $R_{(y \cdot 1)(2 \cdot 1)}^2$ is the partial coefficient of determination of y on x_2 given x_1; this may be seen from a simple rearrangement of (4.20). Thus the condition for reduction of imprecision can be restated as $(1 - R_{(y \cdot 1)(2 \cdot 1)}^2) < (1 - R_{21}^2)$ —or indeed as $R_{(y \cdot 1)(2 \cdot 1)}^2 > R_{21}^2$. In this final version we have the plausible assertion that adding x_2 to the regression of y on x_1 will reduce the imprecision which attaches to the x_1 slope if and only if x_2 is more closely correlated with the regressand y (after controlling for x_1) than it is with the regressor x_1; imprecision being measured by the radex error.

Thus the radex error handles an objection to the index error and should therefore be preferred as a measure of imprecision, particularly when comparing different functions fit to the same data. Fortunately, the simple connection between the index error and the radex error, given in (6.16), means that much of our earlier analysis can be carried over to the new measure with only simple modifications. The reader is invited to undertake this task. We shall note here only a few basic points. The *radex ellipse*, defined as the locus of all combinations of Δ_1 and Δ_2 which raise the sum of squared residuals by just 100 per cent, has the same form as the index ellipse (that is, it is simply a different contour of the same elliptic paraboloid). Indeed, an index ellipse diagram can be reread as a radex ellipse diagram if we simply rescale the coordinate axes: the point originally labeled $(\Delta_1, 0)$ becoming $(\sqrt{\sum e^2}\Delta_1, 0)$, and so forth. Similar remarks apply to the *radex ellipsoid* in the K-regressor case. Also, since $\sum e^2 = (1 - R^2)m_{yy}$, we see from (5.34) that the radex error of a regression coefficient b_k may be expressed as

$$\pi_k^* = \sqrt{[m_{yy}(1 - R_{y \cdot 1, ..., K}^2)]/[m_{kk}(1 - R_{k \cdot 1, ..., k-1, k+1, ..., K}^2)]}. \quad (6.17)$$

A final comment may be in order to avoid possible misunderstanding: The critical distinction between the index error and the radex error is not that the former uses a standard of 1 and the latter uses a standard of 100. Rather it is that the index error uses an absolute standard and the radex error uses a relative standard to measure the seriousness of changes in the sum of squared residuals.

6.6. SUPPLEMENTARY REMARKS

1. Some caution must be exercised in interpreting our claim that the major axis locates that linear combination of regression coefficients which is most precisely estimable. After all, the slope of the major axis reflects only the ratio between the coefficients of the linear combination and not their levels. The index error of a linear combination will depend upon the level of its coefficients not just upon their ratio. [Indeed, it is readily confirmed from (6.7) or (6.8) that doubling all the coefficients of a linear combination will double the index error.] The appropriate interpretation is as follows. Suppose that the level of the coefficients of the linear combination has been normalized by some condition such as $a_1^2 + a_2^2 = h$. Then among all linear combinations whose coefficients meet this condition, the one with the smallest index error is the one which has $-a_1/a_2 = \tan \theta$.

A similar caveat applies to our claim that the orthogonal regression line locates the regressor values for which forecasts are most precisely made. After all, there are an infinity of points on the orthogonal regression line. Of these, the sample mean point $x_{1*} = x_{2*} = 0$ is the one at which forecasts are most precise. Further, it is not hard to see that forecasts for regressor values which lie off the orthogonal regression but close to the origin may be more reliable than forecasts for regressor values which lie on the orthogonal regression but far from the origin. Again the appropriate interpretation is a normalized one. Suppose that distance from the origin is prespecified, $x_1^2 + x_2^2 = h$. Then among all such regressor value pairs, the one at which most reliable forecasts can be made is the one on the orthogonal regression line, namely, (x_{1*}, x_{2*}), where $x_{2*} = (-1/\tan \theta)x_{1*}$.

2. Faced with the problem of fitting $\hat{y} = b_1 x_1 + b_2 x_2$ when x_1 and x_2 are strongly correlated, one might be tempted to "eliminate" or "reduce" the multicollinearity by transformation of the regressors. Specifically when x_1 and x_2 are highly positively correlated, some have proposed to use x_1 and $(x_2 - x_1)$ as regressors rather than x_1 and x_2; the former pair are less correlated than the latter pair. The temptation should be resisted. While the proposed regression $\hat{y} = c_1 x_1 + c_2(x_2 - x_1)$ could yield a reliable estimate for c_1, this is not b_1. The reduced imprecision is obtained not by transforming the regressors but by transforming the problem being answered. Indeed, it is readily confirmed by identifying coefficients in the two relationships that $c_1 = b_1 + b_2$. It should come as no surprise to learn that the sum of regression slopes may be estimated reliably when the regressors are positively correlated.

Tools for investigating the effect of linear transformations of regressors are employed in Goldberger (1964, pp. 185–186, 218–221) and Malinvaud (1966, pp. 44–47). An instructive empirical example on multicollinearity is discussed in Malinvaud (1966, pp. 187–192).

3. Neither the concepts nor the names "radex error," "radex ellipse," or "radex ellipsoid" are in common use. Our adjective "radex" is offered to carry the connotation of "ratio index." The radex concept is a natural second stage in the development of the heuristic approach being taken here. It avoids one deficiency of the index error noted in Section 5.5, item 1. There remains the second deficiency—like the index error, the radex error fails to do justice to the effect of sample size upon reliability of estimation. The classical measure of imprecision, the standard error, does incorporate the sample size explicitly. The radex error turns out to be the product of two components of the standard error. Thus the present approach focuses on two aspects of the classical treatment of imprecision, to the neglect of the others. Once again, this may have some analytical advantage and, at the least, provides a fresh interpretation of part of the traditional computations.

Chapter 7

Stochastic Specification

WE HAVE COME THIS FAR WITH VERY LITTLE IN THE way of a formal statistical model for estimating population regression functions. Indeed, about all that has been assumed is that the PRF is linear. Correspondingly, the only justification for our choice of the LSRF has been the heuristic one that it reproduces in the sample properties that the LPRF has in the population. If we are to provide any more serious justification for the estimation method we must assume something more about the mechanism which generates the sample. After all, if we are to infer something about the PRF from a SRF fitted to a single sample, we must have some idea about how the SRF would vary from sample to sample from the same population. And surely it would vary: If we had drawn a different sample from the same population, we would in general have obtained different values of the *y*'s—even for the same set of *x*'s— and hence we would have computed a different set of *b*'s.

In the classical approach to statistical inference this variability from sample to sample occupies a central position. One assumes that the data at hand constitute a sample from a population; further one assumes some features of the population and of the mechanism which generates samples from it. These assumptions may be referred to as the *stochastic specification* or *statistical model*. On the basis of this stochastic specification it is possible to draw deductions about the probabilities of drawing different samples.

Hence one can deduce something about the probabilities of obtaining various values of sample statistics—such as means, variances, covariances, and regression coefficients. Then, considering the value of the sample statistic computed in a given sample in the light of its sample-to-sample variability, one may infer something about unknown features of the population.

Traditionally, inferences take the form of point estimates of, interval estimates of, or tests of hypotheses about, one or more parameters of the population probability distribution. The choice of a particular sample statistic to estimate a particular population parameter is justified in terms of the sampling distribution of the statistic. It may be established that the distribution has certain desirable properties—for example, that its average value equals the population parameter ("unbiasedness"). Note that the properties are those of the sampling distribution, so that what is justified is the estimation method rather than the value computed in a particular sample; this, is, after all, inherent in the scientific approach in a probabilistic world. In a similiar fashion, the justification for the use of particular sample statistics for interval estimators or hypothesis tests rests upon an evaluation of the sampling distributions of those statistics.

It is, of course, essential to sound inference that the stochastic specification be an appropriate one in the sense that it captures the key features of the process which actually generated the observed data. In this chapter we develop a stochastic specification designed to cover a wide range of economic contexts. As our discussion will indicate, under this stochastic specification the LSRF has certain desirable properties from the point of view of classical statistical inference. This lends support to the heuristic approach developed in the previous chapters. We shall also reconsider the question of precision of estimation from the point of view of classical statistical inference.

7.2. STOCHASTIC SPECIFICATION: ONE-REGRESSOR CASE

We start with the one-regressor case of Section 2.2. Our stochastic specification is as follows. The observed sample, $y_1, x_1, \ldots, y_t, x_t, \ldots, y_T, x_T$, has been generated by a joint probability distribution with the following properties.

The conditional expectation of y_t given $x_1, \ldots, x_t, \ldots, x_T$ is a linear function of x_t only:

$$E(y_t \mid \mathbf{x}) = E(y_t \mid x_t) = \beta_0 + \beta_1 x_t, \tag{7.1}$$

where here and in the remainder of this section $\mathbf{x}$ is shorthand for $(x_1, \ldots, x_t, \ldots, x_T)$. After allowing for the conditional expectation, the conditional distribution of y_t given the x's is the same for all sets of values of the x's; in particular, the variance is constant:

$$E\{[y_t - E(y_t | \mathbf{x})]^2 | \mathbf{x}\} = E[y_t - E(y_t | x_t)]^2 = \sigma^2. \tag{7.2}$$

The T drawings on y are independent; in particular, the covariance between any two drawings is zero:

$$E\{[y_t - E(y_t | \mathbf{x})][y_s - E(y_s | \mathbf{x})] | \mathbf{x}\} = E[y_t - E(y_t | x_t)][y_s - E(y_s | x_s)]$$
$$= 0 \quad \text{for } s \neq t. \tag{7.3}$$

If we define the disturbance $\varepsilon_t = y_t - \beta_0 - \beta_1 x_t$ we may restate these properties as follows, and incidentally clarify the meaning of the phrase "after allowing for the conditional expectation." The conditional distribution of ε_t given $x_1, \ldots, x_t, \ldots, x_T$ is the same for all sets of values of the x's; in particular, the expectation is zero and the variance is constant:

$$E(\varepsilon_t | \mathbf{x}) = E(\varepsilon_t | x_t) = E(\varepsilon_t) = 0, \tag{7.4}$$

$$E(\varepsilon_t^2 | \mathbf{x}) = E(\varepsilon_t^2 | x_t) = E(\varepsilon_t^2) = \sigma^2. \tag{7.5}$$

The T drawings on ε are independent; in particular, the covariance between any two drawings is zero:

$$E(\varepsilon_t \varepsilon_s | \mathbf{x}) = E(\varepsilon_t \varepsilon_s | x_t, x_s) = E(\varepsilon_t \varepsilon_s) = 0 \quad \text{for } s \neq t. \tag{7.6}$$

The disturbance ε's, which are unobserved random variables, should not be confused with the residual e's, which are numbers computable in any sample.

Further, we shall make one rather weak assumption about the distribution of the x's:

$$E\left[1 \bigg/ \sum_{t=1}^{T} (x_t - \bar{x})^2\right] = E(1/m_{11}) \text{ exists}; \tag{7.7}$$

in essence, this says that the x's do vary within samples.

Within this model, we proceed to deduce properties of the sampling distribution of the least-squares estimators; that is, we see what can be said about how the slope and intercept of (2.13) and (2.14) vary from sample to sample. We note two algebraic identities:

$$\sum_{t=1}^{T} (x_t - \bar{x})(y_t - \bar{y}) = \sum_{t=1}^{T} (x_t - \bar{x})y_t - \bar{y} \sum_{t=1}^{T} (x_t - \bar{x})$$

$$= \sum_{t=1}^{T} (x_t - \bar{x})y_t \qquad (7.8)$$

since $\bar{y}$ is constant over the summation and $\sum_{t=1}^{T} (x_t - \bar{x}) = 0$; similarly,

$$\sum_{t=1}^{T} (x_t - \bar{x})^2 = \sum_{t=1}^{T} (x_t - \bar{x})(x_t - \bar{x}) = \sum_{t=1}^{T} (x_t - \bar{x})x_t. \qquad (7.9)$$

Now consider the numerator of the least-squares slope of (2.13). It can be written

$$\sum_{t=1}^{T} (x_t - \bar{x})(y_t - \bar{y}) = \sum_{t=1}^{T} (x_t - \bar{x})y_t$$

$$= \sum_{t=1}^{T} (x_t - \bar{x})(\beta_0 + \beta_1 x_t + \varepsilon_t)$$

$$= \beta_0 \sum_{t=1}^{T} (x_t - \bar{x}) + \beta_1 \sum_{t=1}^{T} (x_t - \bar{x})x_t + \sum_{t=1}^{T} (x_t - \bar{x})\varepsilon_t$$

$$= \beta_1 \sum_{t=1}^{T} (x_t - \bar{x})^2 + \sum_{t=1}^{T} (x_t - \bar{x})\varepsilon_t, \qquad (7.10)$$

where we have used in turn (7.8), the definition of the disturbance, $\sum_{t=1}^{T} (x_t - \bar{x}) = 0$, and (7.9). Dividing (7.10) through by the denominator of (2.13) we find

$$b_{y1} = \beta_1 + \left[\sum_{t=1}^{T} (x_t - \bar{x})\varepsilon_t \right] \Big/ \left[\sum_{t=1}^{T} (x_t - \bar{x})^2 \right], \qquad (7.11)$$

which shows that within the model the least-squares slope estimator is equal to the parameter which it purports to estimate plus a linear combination of the disturbance terms. To examine this linear combination, we define

$$a_t = (x_t - \bar{x}) \Big/ \left[\sum_{t=1}^{T} (x_t - \bar{x})^2 \right] \qquad (t = 1, \ldots, T), \qquad (7.12)$$

and note that in any sample $\sum_{t=1}^{T} a_t = 0$ and $\sum_{t=1}^{T} a_t^2 = 1/m_{11}$. Now (7.11) may be rewritten

$$b_{y1} = \beta_1 + \sum_{t=1}^{T} a_t \varepsilon_t. \tag{7.13}$$

As we go from sample to sample, the least-squares slope will vary about the population slope because the ε's—and possibly also the a's—will vary. What can be deduced about this variation?

We first take the expectation of b_{y1} conditional upon $x_1, \ldots, x_t, \ldots, x_T$ being given; that is, we seek the average value of the least-squares slope across all samples which have the set of x values in common. We find

$$E(b_{y1} | \mathbf{x}) = \beta_1 + E\left[\left(\sum_{t=1}^{T} a_t \varepsilon_t\right) \middle| \mathbf{x}\right] = \beta_1 + \sum_{t=1}^{T} a_t E(\varepsilon_t | \mathbf{x}) = \beta_1, \tag{7.14}$$

using the rule of (2.21), the fact that when the x's are given so are the a's, and (7.4). Next we take the conditional variance of b_{y1} under the same condition. In view of (7.14), this is

$$E[(b_{y1} - \beta_1)^2 | \mathbf{x}] = E\left[\left(\sum_{t=1}^{T} a_t \varepsilon_t\right)^2 \middle| \mathbf{x}\right] = E\left[\left(\sum_{t=1}^{T} \sum_{s=1}^{T} a_t a_s \varepsilon_t \varepsilon_s\right) \middle| \mathbf{x}\right]$$

$$= \sum_{t=1}^{T} \sum_{s=1}^{T} a_t a_s E(\varepsilon_t \varepsilon_s | \mathbf{x}) = \sum_{t=1}^{T} a_t^2 \sigma^2 = \sigma^2/m_{11}, \tag{7.15}$$

where we have used the algebraic formula for the square of a sum, the fact that when the x's are given so are the a's, the rule of (2.21), (7.5), (7.6), and the formula noted above for $\sum_{t=1}^{T} a_t^2$.

We may now readily obtain the mean and variance of the least-squares slope across all samples. From (2.23) we know that any unconditional expectation is the expectation of conditional expectations. From (2.24) we know that if all the conditional expectations have a common value then the unconditional expectation also has that value. Applying these rules to (7.14) and (7.15) we find

$$E(b_{y1}) = E[E(b_{y1} | \mathbf{x})] = E(\beta_1) = \beta_1, \tag{7.16}$$

$$E(b_{y1} - \beta_1)^2 = E[E(b_{y1} - \beta_1)^2 | \mathbf{x}] = E(\sigma^2/m_{11}) = \sigma^2 E(1/m_{11}), \tag{7.17}$$

also using (7.7).

We conclude from (7.16) that in our model, b_{y1} is an *unbiased estimator* of β_1; that is, on the average, the least-squares slope equals the population regression slope. This conclusion provides some justification for using the LSRF to estimate an LPRF. However, there are many other unbiased estimators of the population slope, so that further support is required. What can be shown is that in our model the least-squares slope is the minimum variance linear unbiased estimator of β_1, where linear means linear in $y_1, \ldots, y_T$ and unbiased means unbiased conditionally on every set of values of the x's. It is this concentration of the sampling distribution of the least-squares slope about the population slope that basically justifies our estimation procedure from the point of view of statistical inference within the present model.

The variance formula (7.17) sheds some light on the magnitude of this sample-to-sample variation of the least-squares slope. It therefore suggests a statistical-inference measure of the imprecision of the least-squares slope. We see that the variance of b_{y1} will tend to be large—so that little reliability ought to attach to the least-squares slope computed in a single sample—when σ^2 is large and/or $E(1/m_{11})$ is large. It is not surprising to find that the variance of b_{y1} tends to be high when the disturbance variance is large: When there is considerable variability about the PRF it is quite possible to draw radically different values of y_t for a given value of x_t, and hence to compute radically different SRF's, as we go from sample to sample. Nor is the role of the other term in (7.17) surprising: $E(1/m_{11})$ is simply the average value of $1/m_{11}$, and the discussion of Chapter 5 has suggested why uncertainty should tend to be large when $1/m_{11}$ tends to be large.

To be sure, (7.17) is not quite operational. In practice we are ignorant of σ^2 and $E(1/m_{11})$ and so are unable to compute $E(b_{y1} - \beta_1)^2$ from our sample alone. However, we can obtain an unbiased estimator of it. Let us take the deviations from the LSRF, that is, the sample residuals $e_t = y_t - b_0 - b_{y1}x_t$ $(t = 1, \ldots, T)$, and compute their adjusted mean square: $s^2 = (\sum_{t=1}^{T} e_t^2)/(T - 2)$. We then propose as an estimator of the variance of b_{y1}, the sample statistic

$$s_{b_{y1}}^2 = s^2/m_{11}. \tag{7.18}$$

It can be shown that this estimator is unbiased, that is, that

$$E(s_{b_{y1}}^2) = \sigma^2 E(1/m_{11}).$$

The square root of (7.18), namely,

$$s_{b_{y1}} = \sqrt{1/(T-2)}\sqrt{\sum e^2}\sqrt{1/m_{11}}, \tag{7.19}$$

is called the *standard error of b_{y1}* and is the traditional statistical-inference measure of the imprecision of the least-squares slope. The relationships of this measure to our heuristic measures are clear: The index error is $\sqrt{1/m_{11}}$

and the radex error is $\sqrt{\sum e^2}\sqrt{1/m_{11}}$. It follows from the simplicity of these relations that our discussion in Chapters 5 and 6, which was couched in terms of the index error and radex error, can be reformulated in terms of the standard error. If this is done, the analysis of precision would be placed within a rigorous framework of statistical inference.

In fact, a complete analysis requires additional information about the shape of the sampling distribution of b_{y1}. It is common to add to the specification of the model the assumption that the disturbances are normally distributed, that is, that the conditional distributions of y are normal. With this additional assumption it can be shown that the statistic $(b_{y1} - \beta_1)/s_{b_{y1}}$ has the Student t distribution with $T - 2$ degrees of freedom. Armed with this distributional result one can derive the well-known statistical-inference procedures for interval estimation of, and tests of hypotheses about, the population regression slope. Without reviewing these familiar procedures we may still note the following. When the standard error is large, widely varying hypotheses about the value of β_1 are acceptable and confidence intervals for β_1 are wide. This use of the standard error as a measure of imprecision runs quite parallel to the use of the index error proposed in Section 5.2. Further, a common use of the standard error is for testing the hypothesis that $\beta_1 = 0$, that is, that the PRF is constant. It can be shown that on this null hypothesis the absolute value of the t statistic above can be expressed $\sqrt{R^2/[(1 - R^2)/(T - 2)]}$. Examination of this expression shows that it tends to be large when R^2 and/or T are large. Since large absolute values of a t statistic lead to rejection of the null hypothesis, it is seen, rather reassuringly, that a high R^2 constitutes evidence against the PRF being constant, especially when obtained in a large sample.

Finally, we note that within the present model a justification parallel to that which we have made for the least-squares slope can be made for the least-squares intercept and indeed for the LSRF as a whole.

Example 7.1

Assuming that the present model is applicable to the data of Example 2.4, some results of this section may be illustrated with the results of that example and Examples 4.1 and 5.1. We compute in turn

$$s^2 = \sum e^2/(T - 2) = (304/64)/(6 - 2) = 19/16$$

$$s_{b_{y1}}^2 = s^2/m_{11} = (19/16)/(32/6) \qquad = 57/16^2$$

$$s_{b_{y1}} = \qquad\qquad\qquad\qquad = \sqrt{57}/16$$

We note that $s_{b_{y1}} = \sqrt{1/(T-2)}\sqrt{\sum e^2}\Delta_1^*$, where Δ_1^* is the index error found in Example 5.1. We note also that

$$b_{y1}/s_{b_{y1}} = (5/8)/(\sqrt{57}/16) = 10/\sqrt{57} = \sqrt{(25/82)/[(57/82)/4]}$$

$$= \sqrt{R_y^2 \cdot {}_1/[(1 - R_y^2 \cdot {}_1)/(T-2)]}$$

7.3. STOCHASTIC SPECIFICATION: TWO-REGRESSOR CASE

We proceed to the two-regressor case discussed in Section 3.2. Our stochastic model is specified as follows. The observed sample, y_1, x_{11}, x_{12}, $\ldots, y_t, x_{t1}, x_{t2}, \ldots, y_T, x_{T1}, x_{T2}$, has been generated by a joint probability distribution with the following properties.

The conditional expectation of y_t given $x_{11}, x_{12}, \ldots, x_{t1}, x_{t2}, \ldots, x_{T1}$, x_{T2} is a linear function of x_{t1} and x_{t2} only:

$$E(y_t \mid \mathbf{x}) = E(y_t \mid x_{t1}, x_{t2}) = \beta_0 + \beta_1 x_{t1} + \beta_2 x_{t2}, \qquad (7.20)$$

where here and in the remainder of this section $\mathbf{x}$ is shorthand for $x_{11}, x_{12}, \ldots, x_{t1}, x_{t2}, \ldots, x_{T1}, x_{T2}$. After allowing for the conditional expectation, the conditional distribution of y_t given the x's is the same for all sets of values of the x's; in particular, the variance is constant:

$$E\{[y_t - E(y_t \mid \mathbf{x})]^2 \mid \mathbf{x}\} = E[y_t - E(y_t \mid x_{t1}, x_{t2})]^2 = \sigma^2. \qquad (7.21)$$

The T drawings on y are independent; in particular, the covariance between any two drawings is zero:

$$E\{[y_t - E(y_t \mid \mathbf{x})][y_s - E(y_s \mid \mathbf{x})] \mid \mathbf{x}\} = E\{[y_t - E(y_t \mid x_{t1}, x_{t2})]$$
$$\times [y_s - E(y_s \mid x_{s1}, x_{s2})]\} = 0 \qquad \text{for } s \neq t. \quad (7.22)$$

Again, if we define the disturbance $\varepsilon_t = y_t - \beta_0 - \beta_1 x_{t1} - \beta_2 x_{t2}$ we may restate these properties as follows, and clarify the meaning of the phrase "after allowing for the conditional expectation." The conditional distribution of ε_t given $x_{11}, x_{12}, \ldots, x_{t1}, x_{t2}, \ldots, x_{T1}, x_{T2}$ is the same for all sets of values of the x's; in particular, the expectation is zero and the variance is constant:

$$E(\varepsilon_t \mid \mathbf{x}) = E(\varepsilon_t \mid x_{t1}, x_{t2}) = E(\varepsilon_t) = 0, \qquad (7.23)$$

$$E(\varepsilon_t^2 \mid \mathbf{x}) = E(\varepsilon_t^2 \mid x_{t1}, x_{t2}) = E(\varepsilon_t^2) = \sigma^2. \qquad (7.24)$$

The T drawings on ε are independent; in particular, the covariance between any two drawings is zero:

$$E(\varepsilon_t \varepsilon_s \,|\, \mathbf{x}) = E(\varepsilon_t \varepsilon_s \,|\, x_{t1}, x_{t2}, x_{s1}, x_{s2}) = E(\varepsilon_t \varepsilon_s) = 0 \qquad \text{for } s \neq t. \quad (7.25)$$

Further we require the rather weak assumptions about the distribution of the x's:

$$E[m_{22}/(m_{11}m_{22} - m_{12}^2)], \; E[m_{12}/(m_{11}m_{22} - m_{12}^2)], \text{ and}$$

$$E[m_{11}/(m_{11}m_{22} - m_{12}^2)] \text{ exist}; \quad (7.26)$$

in essence (7.26) says that the regressors vary within samples and not exactly linearly together.

Within this model, some properties of the sampling distributions of the least-squares estimators are readily established; here we shall only state the results, which are special cases of those to be developed in the next section for the general K-regressor case. The least-squares slopes are unbiased estimators of the corresponding population slopes:

$$E(b_{y1 \cdot 2}) = \beta_1, \qquad E(b_{y2 \cdot 1}) = \beta_2. \quad (7.27)$$

They are also minimum variance linear unbiased estimators of those parameters, where linear means linear in $y_1, \ldots, y_T$ and unbiased means unbiased conditionally on every set of values of the x's. The variances and covariance of the least-squares slopes are

$$E(b_{y1 \cdot 2} - \beta_1)^2 = \sigma^2 E[m_{22}/(m_{11}m_{22} - m_{12}^2)],$$

$$E(b_{y2 \cdot 1} - \beta_2)^2 = \sigma^2 E[m_{11}/(m_{11}m_{22} - m_{12}^2)], \quad (7.28)$$

$$E(b_{y1 \cdot 2} - \beta_1)(b_{y2 \cdot 1} - \beta_2) = -\sigma^2 E[m_{12}/(m_{11}m_{22} - m_{12}^2)].$$

Further, the sample statistics

$$s_{b_{y1} \cdot 2}^2 = s^2 [m_{22}/(m_{11}m_{22} - m_{12}^2)],$$

$$s_{b_{y2} \cdot 1}^2 = s^2 [m_{11}/(m_{11}m_{22} - m_{12}^2)], \quad (7.29)$$

$$s_{b_{y1} \cdot 2 b_{y2} \cdot 1} = -s^2 [m_{12}/(m_{11}m_{22} - m_{12}^2)],$$

are unbiased estimators of the corresponding terms in (7.28). In (7.29), $s^2 = (\sum_{t=1}^{T} e_t^2)/(T - 3)$—where $e_t = y_t - b_0 - b_{y1 \cdot 2} x_{t1} - b_{y2 \cdot 1} x_{t2}$—is an unbiased estimator of σ^2.

The square roots of the estimated variances are called the *standard errors* of the least-squares slopes and these are the traditional statistical-inference measures of the imprecision which attaches to $b_{y1 \cdot 2}$ and $b_{y2 \cdot 1}$ respectively. The relationship of these measures to our heuristic measures is again clear. Note further than the variance—and hence the standard error—of $b_{y1 \cdot 2}$ will tend to be large when σ^2 is large and/or $E[m_{22}/(m_{11}m_{22} - m_{12}^2)]$ is large. It is not surprising that imprecision will tend to be large when the disturbance variance is large; nor is the role of the other term surprising. For it is simply the expectation of $1/[m_{11}(1 - R_{21}^2)]$, and the discussion in Chapter 5 has suggested why imprecision should tend to be large when that statistic tends to be large. A parallel interpretation applies to $s_{b_{y2 \cdot 1}}$, of course.

Again a full appreciation of the standard errors requires additional information about the shape of the sampling distributions. It is common to add to the specification of the model the assumption that the disturbances are normally distributed. It can then be shown that each of the statistics $(b_{y1 \cdot 2} - \beta_1)/s_{b_{y1 \cdot 2}}$ and $(b_{y2 \cdot 1} - \beta_2)/s_{b_{y2 \cdot 1}}$ has the Student t distribution with $T - 3$ degrees of freedom; and also that the statistic

$$[(\beta_1 - b_{y1 \cdot 2})^2 m_{11} + (\beta_2 - b_{y2 \cdot 1})^2 m_{22}$$
$$+ 2(\beta_1 - b_{y1 \cdot 2})(\beta_2 - b_{y2 \cdot 1})m_{12}]/2s^2$$

has the Snedecor F distribution with 2 and $T - 3$ degrees of freedom. Application of these results to interval estimation and hypothesis testing would parallel our application of the heuristic measures in Sections 5.3, 6.2, 6.3, and 6.4.

Of particular interest is the intimate relation between the use of the F statistic above and the use of the index ellipse. Consider a joint hypothesis which assigns numerical values to both β_1 and β_2; the formal test of this null hypothesis will lead to acceptance if the sample value of the F statistic is less than or equal to the appropriate critical value, and to rejection otherwise. Thus, where $F_{T-3,\alpha}^2$ denotes the critical value at the 100α per cent level of significance in an F distribution with 2 and $T - 3$ degrees of freedom, the borderline between acceptance and rejection is defined by

$$(\beta_1 - b_{y1 \cdot 2})^2 m_{11} + (\beta_2 - b_{y2 \cdot 1})^2 m_{22}$$
$$+ 2(\beta_1 - b_{y1 \cdot 2})(\beta_2 - b_{2y \cdot 1})m_{12} = 2s^2 F_{T-3,\alpha}^2. \quad (7.30)$$

Any hypothesized pair of values of β_1 and β_2 will then be accepted if it makes the left side of (7.30) less than or equal to the right side, and rejected otherwise. Plotted as a function of the differences between the hypothesized values and the least-squares values, namely, $(\beta_1 - b_{y1 \cdot 2})$ and $(\beta_2 - b_{y2 \cdot 1})$, the equation of (7.30) is simply an ellipse with center at the origin. Indeed as a comparison with (5.6) will show, it is an ellipse *of the very same form as the index ellipse*; note that the differences between the hypothesized values and the least-squares values correspond to Δ_1 and Δ_2, respectively. We may refer to the ellipse defined by (7.30) as a $100(1 - \alpha)$ per cent *confidence ellipse* for β_1 and β_2. For it can be shown that in repeated sampling, such an ellipse covers the point corresponding to the pair of population regression slopes with probability $1 - \alpha$ in the present model.

The gist of the present discussion is that in our statistical-inference context, it is not the index ellipse, but rather a confidence ellipse of the very same form, which is appropriate for discriminating between acceptable and unacceptable deviations of hypothesized values from least-squares values for the pair of population regression slopes. In view of the identity of form of the two ellipses, much of the heuristic discussion of Chapters 5 and 6 carries over with only straightforward modifications to the present model. One way of viewing the relation between the heuristic and the formal procedures is as follows: If we proceed to accept hypotheses which fall in or on the index ellipse and reject those which fall outside it, we will, in the present model, be operating at a level of significance 100α per cent where α is such that $2s^2 F_{T-3,\alpha}^2 = 1$.

As in the one-regressor case, a common use of the distribution results is to test hypotheses about the slope of the PRF. It can be shown that on the null hypothesis that $\beta_1 = 0$, that is, on the hypothesis that the PRF does not vary with respect to the first regressor, the absolute value of first t statistic above reduces to

$$\sqrt{(R_{y \cdot 12}^2 - R_{y \cdot 2}^2)/[(1 - R_{y \cdot 12}^2)/(T - 3)]}.$$

Examination of this expression shows that it tends to be large when the increase in explanation due to introducing the first regressor is large and/or the multiple coefficient of determination is large and/or T is large. Since large absolute values of a t statistic lead to rejection of the null hypothesis, it is seen, rather reassuringly, that a high marginal contribution to R^2 constitutes evidence against the constancy of the PRF with respect to the regressor, particularly when obtained in a large sample and/or when the two regressors together explain a large proportion of the sample variation in y. (A parallel analysis holds, of course, with respect to the null hypothesis $\beta_2 = 0$.) As for the null hypothesis $\beta_1 = \beta_2 = 0$—which asserts that the PRF is constant with respect to both x_1 and x_2—it can be shown that on this null hypothesis the F statistic above reduces to $(R^2_{y \cdot 12}/2)/[(1 - R^2_{y \cdot 12})/(T - 3)]$. Examination of this expression shows that it tends to be large when the multiple coefficient of determination is large and/or T is large. Since large values of an F statistic lead to rejection of the null hypothesis, it follows, reassuringly, that a high multiple coefficient of determination constitutes evidence against the constancy of the PRF, especially when obtained in a large sample.

Example 7.2

Assuming that the present model is applicable to the data of Example 2.5, some results of this section may be illustrated with the results of that example and Examples 5.3 and 4.1. We compute in turn

$$s^2 = \sum e^2/(T-3) = (117/222)/(6-3) = 13/74$$

$$s^2_{b_{y1 \cdot 2}} = (13/74)[(17/6)]/[(32/6)(17/6) - (10/6)^2] = (13)(17)/74^2$$

$$s^2_{b_{y2 \cdot 1}} = (13/74)[(32/6)]/[(32/6)(17/6) - (10/6)^2] = (13)(32)/74^2$$

$$s_{b_{y1 \cdot 2}} = \sqrt{(13)(17)}/74 \qquad s_{b_{y2 \cdot 1}} = \sqrt{(13)(32)}/74$$

We note that $s_{b_{y1 \cdot 2}} = \sqrt{1/(T-3)}\sqrt{\sum e^2}\Delta^*_1$, where Δ^*_1 is the index error found in Example 5.3, and similarly for $s_{b_{y2 \cdot 1}}$. We note also that

$$b_{y2 \cdot 1}/s_{b_{y2 \cdot 1}} = (100/74)/[\sqrt{(13)(32)}/74] = 25/\sqrt{26}$$

$$= \sqrt{[(1400/1517) - (25/82)]/[(117/1517)/3]}$$

$$= \sqrt{(R^2_{y \cdot 12} - R^2_{y \cdot 1})/[(1 - R^2_{y \cdot 12})/(T - 3)]}$$

Figure 7.1 sketches the 95 per cent confidence ellipse for β_1 and β_2. It is of the very same form as the index ellipse of Figure 5.5—indeed the figure can be obtained simply by shrinking the scale of Figure 5.5 by the factor 0.55, since $0.55 \approx 1/\sqrt{3.36}$, where $3.36 = (2)(13/74)(9.55) = 2s^2 F_{3, 0.05}^2$.

7.4. STOCHASTIC SPECIFICATION: K-REGRESSOR CASE

We proceed to the K-regressor case discussed in Section 2.3. To take the most advantage of the compactness of matrix notation, we shall introduce several rather natural definitions: A random matrix is a matrix whose elements are random variables; the probability distribution of a

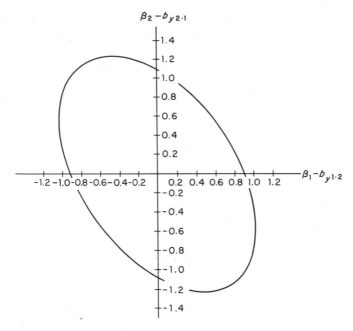

Figure 7.1   95% confidence ellipse for regression coefficients.

random matrix is the joint probability distribution of its elements; two random matrices are stochastically independent if the conditional probability distribution of the first matrix is the same for all values of the second matrix (that is, if the set of elements in the first matrix is stochastically independent of the set of elements in the second matrix); and the expectation of a random matrix is the matrix of expectations of its elements. A vector is, of course, a special case of a matrix, so that the above definitions yield in particular: If $\mathbf{z}$ is a random (column) vector with elements z_i, then the ith element of its expectation vector $E(\mathbf{z})$ is just $E(z_i)$, and the i, j element of its covariance matrix $E[\mathbf{z} - E(\mathbf{z})][\mathbf{z} - E(\mathbf{z})]'$ is the covariance $E[(z_i - E(z_i))][z_j - E(z_j)]$—or variance if $i = j$.

Our stochastic model may now be specified as follows. The observed sample $\mathbf{y}, \mathbf{X}$ has been generated by a joint probability distribution of the random vector $\mathbf{y}$ and the random matrix $\mathbf{X}$ with the following properties. The conditional expectation of $\mathbf{y}$ given $\mathbf{X}$ is

$$E(\mathbf{y} \mid \mathbf{X}) = \mathbf{X}\boldsymbol{\beta}. \tag{7.31}$$

After allowing for the conditional expectation, the conditional distribution of each element of $\mathbf{y}$ given $\mathbf{X}$ is the same for all values of $\mathbf{X}$—in particular, the variances are constant; and the T elements of $\mathbf{y}$ are independent —in particular, the covariance between any two elements of $\mathbf{y}$ is zero:

$$E\{[\mathbf{y} - E(\mathbf{y}|\mathbf{X})][\mathbf{y} - E(\mathbf{y}|\mathbf{X})]' | \mathbf{X}\} = E[\mathbf{y} - E(\mathbf{y}|\mathbf{X})][\mathbf{y} - E(\mathbf{y}|\mathbf{X})]'$$

$$= \sigma^2 \mathbf{I}. \tag{7.32}$$

If we define the disturbance vector $\mathbf{\varepsilon} = \mathbf{y} - \mathbf{X}\mathbf{\beta}$, we may restate those properties and clarify the meaning of the phrase "after allowing for the conditional expectation." The conditional distribution of each element of $\mathbf{\varepsilon}$ given $\mathbf{X}$ is the same for all values of $\mathbf{X}$ and the elements of $\mathbf{\varepsilon}$ are independent; in particular, the disturbance vector has zero expectation and a scalar covariance matrix:

$$E(\mathbf{\varepsilon}|\mathbf{X}) = E(\mathbf{\varepsilon}) = \mathbf{0}, \tag{7.33}$$

$$E(\mathbf{\varepsilon}\mathbf{\varepsilon}'|\mathbf{X}) = E(\mathbf{\varepsilon}\mathbf{\varepsilon}') = \sigma^2 \mathbf{I}. \tag{7.34}$$

Once again the assumption on the distribution of $\mathbf{X}$ itself is rather weak:

$$E[(\mathbf{X}'\mathbf{X})^{-1}] \text{ exists}; \tag{7.35}$$

in essence, this says that the regressors vary within samples and that no set of them vary linearly together.

The properties of the least-squares regression coefficient vector $\mathbf{b}$ are then readily established. Inserting the definition of $\mathbf{\varepsilon}$ into (2.46) we have

$$\mathbf{b} = (\mathbf{X}'\mathbf{X})^{-1}\mathbf{X}'\mathbf{y} = (\mathbf{X}'\mathbf{X})^{-1}\mathbf{X}'(\mathbf{X}\mathbf{\beta} + \mathbf{\varepsilon}) = \mathbf{\beta} + (\mathbf{X}'\mathbf{X})^{-1}\mathbf{X}'\mathbf{\varepsilon}, \tag{7.36}$$

so that the conditional expectation of $\mathbf{b}$ given $\mathbf{X}$—and hence given $(\mathbf{X}'\mathbf{X})^{-1}\mathbf{X}'$—is

$$E(\mathbf{b}|\mathbf{X}) = \mathbf{\beta} + (\mathbf{X}'\mathbf{X})^{-1}\mathbf{X}'E(\mathbf{\varepsilon}|\mathbf{X}) = \mathbf{\beta}, \tag{7.37}$$

using the rule of (2.21) and (7.33). In view of this, we may obtain the conditional covariance matrix of $\mathbf{b}$ given $\mathbf{X}$ as

$$E[(\mathbf{b} - \mathbf{\beta})(\mathbf{b} - \mathbf{\beta})' | \mathbf{X}] = E[(\mathbf{X}'\mathbf{X})^{-1}\mathbf{X}'\mathbf{\varepsilon}\mathbf{\varepsilon}'\mathbf{X}(\mathbf{X}'\mathbf{X})^{-1} | \mathbf{X}]$$

$$= (\mathbf{X}'\mathbf{X})^{-1}\mathbf{X}'E(\mathbf{\varepsilon}\mathbf{\varepsilon}' | \mathbf{X})\mathbf{X}(\mathbf{X}'\mathbf{X})^{-1}$$

$$= (\mathbf{X}'\mathbf{X})^{-1}\mathbf{X}'\sigma^2\mathbf{I}\mathbf{X}(\mathbf{X}'\mathbf{X})^{-1}$$

$$= \sigma^2(\mathbf{X}'\mathbf{X})^{-1} \tag{7.38}$$

using the rule of (2.21) and (7.34). And now the unconditional expectation vector and covariance matrix of **b** are seen to be

$$E(\mathbf{b}) = E[E(\mathbf{b}\,|\,\mathbf{X})] = \boldsymbol{\beta}, \tag{7.39}$$

$$E(\mathbf{b} - \boldsymbol{\beta})(\mathbf{b} - \boldsymbol{\beta})' = E\{E[(\mathbf{b} - \boldsymbol{\beta})(\mathbf{b} - \boldsymbol{\beta})'\,|\,\mathbf{X}]\} = \sigma^2 E[(\mathbf{X}'\mathbf{X})^{-1}] \tag{7.40}$$

using the rules (2.23) and (2.24), and (7.35).

We conclude that in our model the least-squares regression coefficient vector is unbiased; it can be shown that it is indeed the minimum variance linear unbiased estimator of the population regression coefficient vector, where linear means linear in **y** and unbiased means unbiased conditionally on every value of **X**.

It will be clear that (7.40) is not operational, since neither σ^2 nor $E[(\mathbf{X}'\mathbf{X})^{-1}]$ will be known in practice; but again an unbiased estimator of the covariance matrix can be obtained from a single sample. Define the residual vector $\mathbf{e} = \mathbf{y} - \mathbf{Xb}$ and the adjusted mean residual sum of squares $s^2 = \mathbf{e}'\mathbf{e}/(T - K - 1)$. It can be shown that s^2 is an unbiased estimator of σ^2 conditionally on every value of **X** and hence unconditionally; that is, $E(s^2\,|\,\mathbf{X}) = E(s^2) = \sigma^2$. Further, it is obvious that $(\mathbf{X}'\mathbf{X})^{-1}$ is an unbiased estimator of its own expectation. Thus the estimated covariance matrix, namely,

$$\mathbf{S}_{bb} = s^2(\mathbf{X}'\mathbf{X})^{-1}, \tag{7.41}$$

is an unbiased estimator of $E(\mathbf{b} - \boldsymbol{\beta})(\mathbf{b} - \boldsymbol{\beta})'$; that is, $E(\mathbf{S}_{bb}) = E\{E[s^2(\mathbf{X}'\mathbf{X})^{-1}]\,|\,\mathbf{X}\} = E\{(\mathbf{X}'\mathbf{X})^{-1}E(s^2\,|\,\mathbf{X})\} = \sigma^2 E[(\mathbf{X}'\mathbf{X})^{-1}]$. The square root of the kth diagonal element of $\mathbf{S}_{bb}$ is called the standard error of b_k; this statistical-inference measure of the imprecision of b_k is again proportional to the index error developed in Section 5.4 (and also to the radex error). The off-diagonal elements of $\mathbf{S}_{bb}$ are related to the compensating variations discussed in Chapter 5. Again, the discussion presented there suggests the interpretation of the roles played by σ^2 and $(\mathbf{X}'\mathbf{X})^{-1}$ in measuring the imprecision of least-squares regression coefficients.

If the disturbance vector $\boldsymbol{\varepsilon}$ is assumed to be normally distributed, then it can be shown that each of the statistics $(b_k - \beta_k)/s_{b_k}$ has the Student t distribution with $T - K - 1$ degrees of freedom; that the statistic $[(\boldsymbol{\beta} - \mathbf{b})'\mathbf{X}'\mathbf{X}(\boldsymbol{\beta} - \mathbf{b})]/[(1 + K)s^2]$ has the Snedecor F distribution with $(1 + K)$ and $(T - K - 1)$ degrees of freedom; and that where $\boldsymbol{\beta}_2$ is any $(K - H) \times 1$ subvector of $\boldsymbol{\beta}$, $\mathbf{b}_2$ the corresponding $(K - H) \times 1$ subvector of **b**, and $\mathbf{M}^{22}$ the corresponding $(K - H) \times (K - H)$ submatrix of

$(\mathbf{X}'\mathbf{X})^{-1}$, the statistic $[(\boldsymbol{\beta}_2 - \mathbf{b}_2)'(\mathbf{M}^{22})^{-1}(\boldsymbol{\beta}_2 - \mathbf{b}_2)]/[(K - H)s^2]$ has the Snedecor F distribution with $K - H$ and $T - K - 1$ degrees of freedom.

From these distributional results one can derive the well-known procedures for constructing interval estimates of, and testing hypotheses about, the population regression coefficients. The relation between these formal statistical procedures and the heuristic procedures of Chapters 5 and 6, and the relations between the test statistics and coefficients of determination, are simply extensions of those considered in the two-regressor case of Section 7.3. We note only the following explicitly. On the hypothesis that the vector of regression coefficients $\boldsymbol{\beta}_2$ is zero—that is, that the PRF does not vary with those regressors—the second F statistic above reduces to $[(R_{y.12}^2 - R_{y.1}^2)/(K - H)]/[(1 - R_{y.12}^2)/(T - K - 1)]$. Examination of this expression shows that it tends to be large when $(R_{y.12}^2 - R_{y.1}^2)$ is large and/or $(K - H)$ is small and/or $R_{y.12}^2$ is large and/or T is large. Large values of an F statistic lead to rejection of the null hypothesis, so that, reassuringly enough, a high marginal contribution to R^2 constitutes evidence against the constancy of the PRF with respect to a subset of regressors, especially when obtained with few regressors in a large sample with considerable variation explained by the full set of regressors.

7.5. SUPPLEMENTARY REMARKS

1. The stochastic specification proposed in this chapter is an extension of the classical linear regression model. The extension consists of allowing the regressors to be stochastic rather than requiring that their values be fixed in repeated samples. This means that the proposed specification is appropriate in a wider range of economic contexts than is the classical linear regression model. Nevertheless, the present specification does require that the regressors be distributed independently of the disturbances. (The force of this requirement is to insist that the classical model holds conditionally on the regressors being given). There remain many economic contexts in which the proposed model is not applicable. Indeed, one may say that the theory of econometrics is concerned with (a) formulating alternative stochastic specifications which have weaker assumptions and hence are more widely applicable in economics, and (b) developing methods of estimation which are appropriate under those weaker specifications. It turns out that the least-squares estimators have some desirable properties under assumptions somewhat weaker than used here; in any event, a study of their properties is usually a convenient starting point in the search for appropriate estimators.

For more detailed presentations of various aspects of the present specifications—including derivations of results for which we have simply said "it can be shown"—see Goldberger (1964, pp. 267–272), Johnston (1963, pp. 25–29), and Malinvaud (1966, pp. 12–15, 94–97). Further, Johnston (1963, pp. 145–147) concisely indicates the reasons why the present kind of specification may be inapplicable in many economic contexts.

2. We have here attempted to deepen the previous discussion of this book by placing the problem of estimating LPRF's within a well-defined statistical-inference framework. The properties thus established for least-squares estimates of regression coefficients and for associated statistics provide some justification for the least-squares computations in empirical research, assuming that the model is appropriate to the data being analyzed.

Many of the results of the preceding chapters now deserve reinterpretation from the point of view of the present model. We have already sketched some of this reinterpretation for the measures of precision. The reader is invited to complete the task. As one illustration we consider the formulas developed in Chapter 3 for the effect upon regression slopes of adding or deleting regressors.

Suppose that the PRF is known to be linear in the conditioning variables: $E(y \mid x) = \beta_0 + \beta_1 x_1 + \beta_2 x_2$, and that the other specifications of Section 7.3 also hold. Then the slopes in the multiple regressions, $b_{y1 \cdot 2}$ and $b_{y2 \cdot 1}$ are unbiased estimators of β_1 and β_2, respectively. But the slopes from the simple regressions, b_{y1} and b_{y2}, are in general *biased* estimators of those parameters. This can be seen as follows: Taking expectations in (3.15) gives

$$E(b_{y1}) = E(b_{y1 \cdot 2}) + E(b_{21})E(b_{y2 \cdot 1}) = \beta_1 + \beta_{21}\beta_2, \qquad (7.42)$$

where β_{21} simply denotes $E(b_{21})$ and we were enabled to write the expectation of a product as the product of expectations because of the independence between disturbances and regressors. Now $\beta_1 + \beta_{21}\beta_2$ differs from β_1 unless either $\beta_{21} = 0$ (the population covariance between the two regressors is zero) or $\beta_2 = 0$ (x_2 does not belong in the PRF). This emphasizes the dangers of omitting relevant variables in estimating regression functions.

Chapter 8   Functional Forms

8.1. INTRODUCTION

IN CHAPTER 2 WE INTRODUCED THE CONCEPT OF A stochastic relationship: a relationship in which the values of one set of variables determine the (conditional) probability distribution of another variable. We soon decided to concentrate on only one aspect of such a relationship—the conditional expectation—to the neglect of other aspects such as the conditional variance and the shape of the distributions. Presumably, this neglect is justified, since in most empirical research the main feature of interest is precisely the conditional expectation function (=PRF), which describes how the average value of the one variable varies as a function of the others.

However, apart from a few passages in Chapter 2, we have been analyzing only *linear* population regression functions, to the neglect of alternative functional forms. This neglect would seem to be less justifiable. After all, there are many contexts in which there are good theoretical and/or empirical reasons to expect that the conditional expectation of one variable varies in a *nonlinear* way with the values of the others. For example, the classical theory of the firm specifies a U-shaped cost function, cross sections of households display saturation levels of expenditure on particular commodities, and so forth. It would certainly appear that a nonlinear PRF is required when such nonlinear relationships are given a stochastic formulation.

Does this imply that all our analysis and estimation technique is limited to the case of linear relationships? The purpose of this chapter is to demonstrate that the answer to this question is in the negative. As we shall see, the expression "linear regression analysis" is a somewhat misleading description of the methods developed in this book; it fails to do justice to the flexibility of the theory and estimation technique.

Before proceeding to the main argument, it is worthwhile to clarify a few mathematical concepts. Consider first the function of one variable $Y = f(X)$. We say that it is *linear in* X if and only if dY/dX does not involve X, that is, if and only if $d(dY/dX)/dX = 0$, that is, if and only if the effect of a given change in X does not depend on the level of X. Consider next the function of two variables $Y = f(X_1, X_2)$. We say that it is linear in X_1 if and only if $\partial Y/\partial X_1$ does not involve X_1, that is, if and only if $\partial(\partial Y/\partial X_1)/\partial X_1 = 0$, that is, if and only if the effect of a given change in X_1 does not depend on the level of X_1. Similarly, we say that it is linear in X_2 if and only if $\partial Y/\partial X_2$ does not involve X_2, that is, if and only if $\partial(\partial Y/\partial X_2)/\partial X_2 = 0$, that is, if and only if the effect of a given change in X_2 does not depend on the level of X_2. Thus essentially the same concept of linearity applies whether there are one or more independent variables. But a new concept also applies: We say that $Y = f(X_1, X_2)$ is *additive in* X_1, X_2 if and only if $\partial Y/\partial X_1$ does not involve X_2 and $\partial Y/\partial X_2$ does not involve X_1, that is, if and only if $\partial(\partial Y/\partial X_1)/\partial X_2 = 0 = \partial(\partial Y/\partial X_2)/\partial X_1$, that is, if and only if the effect of a given change in each of the independent variables does not depend upon the level of the other. Additivity is an appropriate name for this feature since it means that the combined effect of given changes in both independent variables can be obtained by adding together the separately computed effects of the given changes in each of them. The extension to the case of a function of many variables is straightforward; also, the concepts can be formulated in terms of finite changes to cover the case where derivatives are not defined.

To clarify the concepts of linearity and additivity the reader is invited to confirm the features given for functions (a) to (h) in the following tabulation.

Function $f(X_1, X_2)$	$\partial f/\partial X_1$	$\partial f/\partial X_2$	Linear in X_1?	Linear in X_2?	Additive in X_1, X_2?
(a) $X_1^2 + X_2^2 + 2X_1X_2$	$2(X_1 + X_2)$	$2(X_1 + X_2)$	No	No	No
(b) X_2/X_1	$-X_2/X_1^2$	$1/X_1$	No	Yes	No
(c) $a_1X_1^2 + a_2X_2$	$2a_1X_1$	a_2	No	Yes	Yes
(d) $(X_1 - a_1)^2 + (X_2 - a_2)^2$	$2(X_1 - a_1)$	$2(X_2 - a_2)$	No	No	Yes
(e) $a_0 + a_1X_1X_2^2 + a_2 \log X_2$	$a_1X_2^2$	$2a_1X_1X_2 + a_2/X_2$	Yes	No	No
(f) $a_0 + a_1X_1X_2$	a_1X_2	a_1X_1	Yes	Yes	No
(g) $a_0 + a_1X_1 + a_2 \log X_2$	a_1	a_2/X_2	Yes	No	Yes
(h) $a_0 + a_1X_1 + a_2X_2$	a_1	a_2	Yes	Yes	Yes

This set of functions illustrates several points. First, all combinations of the properties of linearity and additivity are in fact possible. Second, the regression functions which we have been calling "linear" are, strictly speaking, "linear and additive"—see function (h); however, we shall retain the shorter description, which is the standard one. Third, and most important for our present purposes, it is often possible to express a nonlinear function as a linear one by a transformation of variables. For example, function (b) is equivalent to (b*) $X_2 X_1^*$, where $X_1^* = 1/X_1$; (b*) is linear in its arguments (although still not additive in them). Indeed, it is often possible to express a nonlinear nonadditive function as a linear additive one by appropriate transformation of variables. For example, function (a) is equivalent to (a*) $X_1^* + X_2^* + 2X_3^*$, where $X_1^* = X_1^2$, $X_2^* = X_2^2$, and $X_3^* = X_1 X_2$; (a*) is linear and additive in its arguments. This suggests an approach to estimating nonlinear PRF's: Transform them into linear ones. The feasibility of this approach is explored in this chapter.

The key idea in the exploration is the following. The theory and technique of linear regression require that the conditional expectation of *the regressand y be "linear"* (i.e., linear and additive) *in the regressors* $x_1, \ldots, x_K$. But these variables need not be the original variables of interest in our problem. Suppose that we are in fact interested in the expectation of a *dependent* (=to be explained) *variable Y*, conditional upon the values of a set of *conditioning* (= independent = explanatory) *variables* $X_1, \ldots, X_J$. Then it is possible that some function of Y, say, y, has a conditional expectation which is "linear" (i.e., linear and additive) in a set of functions of the X's, say, $x_1, \ldots, x_K$. The parameters of this latter conditional expectation will, of course, be related to the parameters of the original conditional expectation of interest. Further, conditioning on the X's amounts to conditioning on the x's—since "given the X's" implies "given the x's." Thus it may well be possible to apply linear regression to a wide variety of nonlinear problems.

8.2.   CONVENTIONAL NONLINEAR FUNCTIONAL FORMS

A few familiar cases will begin to illustrate how nonlinear regression problems can be transformed into linear ones.

Suppose that we know, or are willing to assume, that the PRF has the *reciprocal* form

$$E(y \mid X) = \beta_0 + \beta_1/X. \tag{8.1}$$

This form, with $\beta_0 > 0$, $\beta_1 < 0$, becomes positive at $X = (-\beta_1/\beta_0) > 0$ and has an asymptote at $E(y \mid X) = \beta_0$. Therefore it has been used in Engel curve studies of the relationship between expenditures on a particular commodity, y, and income X, where economic theory suggests that there is a minimum income below which none of this commodity will be purchased, and a maximum (=saturation) level of expenditures on this commodity which will not be exceeded regardless of income; see Prais and Houthakker (1955, pp. 82–84). In any event, can we estimate this PRF on the basis of a sample of observations y_t, X_t ($t = 1, \ldots, T$)? Clearly, we can. We define the regressor $x = 1/X$ and rewrite (8.1) as

$$E(y \mid X) = \beta_0 + \beta_1 x, \tag{8.2}$$

which is also $E(y \mid x)$, and note that our original sample also provides us with a transformed sample y_t, x_t ($t = 1, \ldots, T$). Hence all the analysis and technique of the preceding chapters is applicable and we can estimate the parameters of (8.2)—that is, of (8.1)—by taking the regression of y on x.

For another example, suppose that we know, or are willing to assume, that the PRF has the *quadratic* form

$$E(y \mid X) = \beta_0 + \beta_1 X + \beta_2 X^2. \tag{8.3}$$

This form, with $\beta_0 > 0$, $\beta_1 < 0$, and $\beta_2 > 0$, falls and then rises as X rises. Therefore, it has been used in industry studies of the relationship between marginal cost, y, and output, X_2, where economic theory suggests that the relationship is U-shaped; for variants, see Johnston (1960). Given a sample of observations y_t, X_t ($t = 1, \ldots, T$) we can estimate this PRF as follows. We define the regressors $x_1 = X$ and $x_2 = X^2$, rewrite (8.3) as

$$E(y \mid X) = \beta_0 + \beta_1 x_1 + \beta_2 x_2, \tag{8.4}$$

which is also $E(y \mid x_1, x_2)$, and recognize that our original sample also provides us with a transformed sample y_t, x_{t1}, x_{t2} ($t = 1, \ldots, T$). Again all the previous analysis is applicable and we can estimate the parameters of (8.4)—that is, of (8.3)—by taking the multiple regression of y on x_1 and x_2. This clearly extends to higher-order polynomials since $\sum_{k=1}^{K} \beta_k X^k = \sum_{k=1}^{K} \beta_k x_k$, where $x_k = X^k$.

That there are limits to this transformation device is shown by the following example. Suppose that we know, or are willing to assume, that the PRF has the form

$$E(y \mid X) = \beta_0 + \frac{\beta_1}{X - \beta_2}. \tag{8.5}$$

This form, with $\beta_0 > 0$, $\beta_1 > 0$, and $\beta_2 > 0$, defined for $X \geqslant \beta_2$, falls from infinity toward an asymptote at β_0 as X rises from β_2 to infinity. Therefore, it may be used in demand-for-money studies, where economic theory (Keynes' liquidity preference) suggests that the relationship between cash holdings, y, and interest rates, X, has some such shape; see Klein and Goldberger (1955, pp. 23–26) for a variant of this. It is clear that we can define $x = 1/(X - \beta_2)$ and rewrite (8.5) as

$$E(y \mid X) = \beta_0 + \beta_1 x, \tag{8.6}$$

which is also $E(y \mid x)$. But now a sample of observations on the original variables y_t, X_t $(t = 1, \ldots, T)$ *does not* provide us with observations on the transformed variables: Since β_2 is unknown, we cannot compute the $x_t = 1/(X_t - \beta_2)$. Thus the linear regression analysis and estimation technique are not directly applicable.

The lesson to be learned from these examples is that the linear regression approach does not require that the PRF be linear in the (original) conditioning variables. If it is linear in *known* functions of the conditioning variables, then taking these known functions as the regressors gives a linear PRF which can be estimated by the linear SRF. The difficulty in the last example was that the required regressors were not known functions of the original conditioning variables; that is, they could not be computed from the sample.

While the examples have referred to the case of a single conditioning variable, it is clear that the transformation device can also be used to obtain a linear model when there are several conditioning examples. One example should suffice to make this point. Consider the following function which has been proposed for the analysis of cross-section household behavior:

$$E(y \mid X_1, X_2, X_3) = \beta_1 X_1 + \beta_2(X_1 \log X_1) + \beta_3(X_1 \log X_2)$$
$$+ \beta_4 X_3 + \beta_5(X_1 X_3), \tag{8.7}$$

where y = savings, X_1 = disposable income, X_2 = family size, and X_3 = beginning-of-year liquid assets; for a variant, see Klein (1951). This highly nonlinear function can be brought into our linear model as

$$E(y \mid X_1, X_2, X_3) = \beta_1 x_1 + \beta_2 x_2 + \beta_3 x_3 + \beta_4 x_4 + \beta_5 x_5, \tag{8.8}$$

if we define the regressors $x_1 = X_1$, $x_2 = X_1 \log X_1$, $x_3 = X_1 \log X_2$, $x_4 = X_3$, and $x_5 = X_1 X_3$.

In summary, we have seen that if we are interested in $E(y \mid X_1, \ldots, X_J) = g(X_1, \ldots, X_J)$, and if the function $g(X_1, \ldots, X_J)$ can be expressed as $\beta_0 + \sum_{k=1}^{K} \beta_k x_k$, where the $x_k = h_k(X_1, \ldots, X_J)$ $(k = 1, \ldots, K)$ are known functions, then the sample $y_t, X_{t1}, \ldots, X_{tJ}$ $(t = 1, \ldots, T)$ can be viewed as a sample $y_t, x_{t1}, \ldots, x_{tK}$ $(t = 1, \ldots, T)$, and linear regression can be applied to the transformed function.

8.3. DUMMY-VARIABLE FUNCTIONAL FORMS

In Section 8.2 we considered only cases where the conditional expectation of the dependent variable was known, or assumed, to be a conventional (analytic) function of the conditioning variables. We assumed that the shape of this function was known and that the sample was being used to estimate some unknown parameters. While such models have been in very frequent use in economic theory and research for many years, they do not exhaust the possibilities. Particularly in recent years it has become very common to utilize unconventional functional forms based upon " dummy variables." We consider several examples to give a feel for this approach which, as we shall see, makes linear regression analysis very flexible indeed.

In Chapter 2 we very briefly mentioned the cell-mean approach to estimating a PRF, but soon dropped it in favor of the linear regression model. We now show that even this cell-mean approach can be formulated as a linear regression problem. Suppose we know, or are willing to assume, that the PRF has the *point* form

$$E(y \mid X) = \begin{cases} \beta_1 & \text{if } X = X^{(1)} \\ \vdots & \vdots \\ \beta_k & \text{if } X = X^{(k)} \\ \vdots & \vdots \\ \beta_K & \text{if } X = X^{(K)}, \end{cases} \tag{8.9}$$

where the $X^{(k)}$ $(k = 1, \ldots, K)$ denote the K distinct possible values of X. This is a very mild assumption, indeed, since it allows the conditional expectation to vary freely as X varies. No smoothness is imposed, as would be done by conventional forms. Now (8.9) can be rewritten in a linear form,

$$E(y \mid X) = \beta_1 x_1 + \cdots + \beta_k x_k + \cdots + \beta_K x_K, \tag{8.10}$$

if we simply define the dummy-variable regressors

$$x_k = \begin{cases} 1 & \text{if } X = X^{(k)}, \\ 0 & \text{if } X \neq X^{(k)}. \end{cases}$$

Clearly our original sample y_t, X_t $(t = 1, \ldots, T)$ can be viewed as a sample $y_t, x_{t1}, \ldots, x_{tK}$ $(t = 1, \ldots, T)$, and linear regression can be applied to estimate the β's in (8.10)—which are the same as those of (8.9). Note that for each observation one and only one x has value 1; the others are all 0.

It is not hard to see that the least-squares estimates of the β's are simply the cell means of y—the set of average values of y with the averaging being done separately for each value of X. Without any loss of generality we can reorder the observations so that the first T_1 have $X = X^{(1)}$, the next T_2 have $X = X^{(2)}, \ldots$, the last T_K have $X = X^{(K)}$; thus we have K groups or cells with the number of observations in the kth cell being T_k, and with $\sum_{k=1}^{K} T_k = T$. Once this is done, it becomes apparent that in view of the very special structure of the x's, the moments which enter the least-squares calculation also have a very special structure:

$$\sum_{t=1}^{T} x_{tk}^2 = (0^2 + \cdots + 0^2) + \cdots + (1^2 + \cdots + 1^2) + \cdots$$

$$+ (0^2 + \cdots + 0^2) = T_k,$$

$$\sum_{t=1}^{T} x_{tk} x_{tj} = 0 \quad \text{if } j \neq k, \tag{8.11}$$

$$\sum_{t=1}^{T} x_{tk} y_t = \sum_{t=T_1+\cdots+T_{k-1}+1}^{T_1+\cdots+T_k} y_t = \sum^{(k)} y,$$

where the last summation sign is shorthand for the next-to-last one, so that $\sum^{(k)} y$ is simply the sum of y's in the kth cell. Then the normal equations have the very simple structure

$$T_1 b_1 + \cdots + 0 b_k + \cdots + 0 b_K = \sum^{(1)} y$$

$$\vdots \qquad\qquad \vdots \qquad\qquad \vdots \qquad \vdots$$

$$0 b_1 + \cdots + T_k b_k + \cdots + 0 b_K = \sum^{(k)} y \tag{8.12}$$

$$\vdots \qquad\qquad \vdots \qquad\qquad \vdots \qquad \vdots$$

$$0 b_1 + \cdots + 0 b_k + \cdots + T_K b_K = \sum^{(K)} y$$

and hence the very simple solutions

$$b_1 = \sum^{(1)} y/T_1, \ldots, b_k = \sum^{(k)} y/T_k, \ldots, b_K = \sum^{(K)} y/T_K, \qquad (8.13)$$

which are in fact the cell means, $\bar{y}_k$ $(k = 1, \ldots, K)$, say. This result—that in the present model the conditional expectation function is estimated by the corresponding sample means—should come as no surprise. After all, the mildness of the specification (8.9) implies that observations outside a cell contain no information relevant to estimating the conditional expectation for that cell.

We have just seen that the highly nonlinear function (8.9) can be put in a very simple linear form, again by transformation of variables. We can also see why this approach is unlikely to be feasible in practice unless there are only a few possible values of X and many observations at each of these values. For the index error of b_k is obviously $\sqrt{1/T_k}$, which will be large if T_k, the number of observations in cell k, is small. Similar remarks apply for the radex and standard errors. In brief, reliable estimates of a mean cannot be obtained from a small sample, and in the present model each cell is effectively a sample by itself.

We proceed to another example. The specification (8.9) requires that the conditioning variable be discrete. While some conditioning variables —for example, family size—are of this type, most are continuous. Then an adaptation of (8.9) may be relevant. If we know or are willing to assume that the PRF has the *step* form

$$E(y \mid X) = \begin{cases} \beta_1 & \text{if } X^{(0)} \leq X < X^{(1)} \\ \vdots & \vdots \\ \beta_k & \text{if } X^{(k-1)} \leq X < X^{(k)} \\ \vdots & \vdots \\ \beta_K & \text{if } X^{(K-1)} \leq X < X^{(K)}, \end{cases} \qquad (8.14)$$

so that the $X^{(k)}$'s mark off intervals instead of being discrete values themselves, then the corresponding definition of dummy-variable regressors will lead to a linear model. The cells are then defined by intervals rather than discrete values, so that we may have more observations per cell. Note that (8.14) amounts to saying that the PRF is a step function which remains constant over intervals but jumps (up or down, by an amount to be estimated) from interval to interval. In certain contexts this may well be more appropriate than assuming that the PRF has some particular smooth analytic form.

Incidentally, the present dummy-variable devices should not be confused with a coding device which regresses y on x^*, where the regressor x^* is a coded value of X: say, $x^* = 1$ if $X = X^{(1)}$, $x^* = 2$ if $X = X^{(2)}, \ldots,$ $x^* = K$ if $X = X^{(K)}$. This latter regression is appropriate for estimating $E(y \mid X) = \beta_0^* + \beta_1^* x_1^*$, which is more restrictive than (8.9): It specifies that the conditional expectation changes by β_1^* as we move from any discrete value of X to the next, while in (8.9) this change, $\beta_{k+1} - \beta_k$, may well be different for different values of X.

The dummy-variable device extends to the case of more than one conditioning variable. Suppose, for example, that we know, or are willing to assume, that the PRF takes the form

$$E(y \mid X_1, X_2) = \beta_{kj} \qquad \text{if } X_1 = X_1^{(k)} \text{ and } X_2 = X_2^{(j)}$$

$$(k = 1, \ldots, K; j = 1, \ldots, J) \quad (8.15)$$

where the $X_1^{(k)}$ and $X_2^{(j)}$ denote the K and J distinct possible values of X_1 and X_2, respectively. (An extension to the case of intervals is again possible.) Here again we have a very mild assumption, since it allows the conditional expectation to vary freely as X_1 and/or X_2 vary; the PRF can have $K \times J$ distinct values. This effectively divides the observations into $K \times J$ cells, leads to the definition of the $K \times J$ dummy-variable regressors

$$x_{kj} = \begin{cases} 1 & \text{if } X_1 = X_1^{(k)} \text{ and } X_2 = X_2^{(j)}, \\ 0 & \text{otherwise,} \end{cases}$$

and to estimation of the linear PRF,

$$E(y \mid X_1, X_2) = \sum_{k=1}^{K} \sum_{j=1}^{J} \beta_{kj} x_{kj}. \qquad (8.16)$$

It can readily be confirmed that the least-squares estimates are simply the cell means, $b_{kj} = \sum^{(kj)} y / T_{kj} = \bar{y}_{kj}$, where T_{kj} denotes the number of observations in the k, j cell.

It is important to distinguish (8.16) from the following. Suppose we know or are willing to assume that the PRF has the form

$$E(y \mid X_1, X_2) = \beta_k + \gamma_j \qquad \text{if } X_1 = X_1^{(k)} \text{ and } X_2 = X_2^{(j)}$$

$$(k = 1, \ldots, K; j = 1, \ldots, J). \quad (8.17)$$

This is a rather stronger assumption. While the conditional expectation can still take on $K \times J$ distinct values, these values are not unconnected; (8.17) has only $K + J$, and not $K \times J$, parameters. In (8.17) the difference between the conditional expectations in the two adjacent cells $k + 1, j$ and k, j—namely, $(\beta_{k+1} + \gamma_j) - (\beta_k + \gamma_j) = \beta_{k+1} - \beta_k$—is the same for all j; while in (8.15) this difference—namely, $\beta_{k+1,j} - \beta_{kj}$—can differ for different j. In the terms of Section 8.1, (8.17) has additive effects while (8.15) does not. In the language of analysis of variance, (8.17) rules out *interactions* while (8.15) permits them.

The linear version of (8.17) would appear to be

$$E(y \mid X_1, X_2) = \sum_{k=1}^{K} \beta_k x_k + \sum_{j=1}^{J} \gamma_j z_j, \qquad (8.18)$$

where the dummy-variable regressors are defined as

$$x_k = \begin{cases} 1 & \text{if } X_1 = X_1^{(k)} \\ 0 & \text{if } X_1 \neq X_1^{(k)} \end{cases} \qquad (k = 1, \ldots, K),$$

$$z_j = \begin{cases} 1 & \text{if } X_2 = X_2^{(j)} \\ 0 & \text{if } X_2 \neq X_2^{(j)} \end{cases} \qquad (j = 1, \ldots, J).$$

This is essentially correct, but a slight modification is necessary to avoid exact multicollinearity among the regressors; see, for example, Goldberger (1964, pp. 218–224). In any event, the conditional expectations will no longer be estimated by the cell means; there are, after all, only $K + J$ free parameters but $K \times J$ cell means. This corresponds to the fact that the additivity assumption imparts a smoothness to the PRF, so that observations in other cells do affect the estimate of the conditional expectation in any single cell.

While our illustrations have involved ordinary, quantitative (=numerical) variables, nothing in the present approach prevents the conditioning variables from being qualitative (= attribute = categorical) variables. For example, we may think of the income of individuals as depending upon their occupation in the sense that the income distribution is different in different occupations. Thus if we distinguish K different occupations, label them arbitrarily $1, \ldots, k, \ldots, K$, we may be interested in the PRF of income y,

$$E(y \mid \text{occupation}) = \beta_k \text{ for occupation } k \qquad (k = 1, \ldots, K). \quad (8.19)$$

The linear version of this is

$$E(y \mid \text{occupation}) = \beta_1 x_1 + \cdots + \beta_K x_K, \qquad (8.20)$$

where the regressors are defined by

$$x_k = \begin{cases} 1 & \text{in occupation } k \\ 0 & \text{in other occupations} \end{cases} \qquad (k = 1, \ldots, K).$$

Note in particular that there is nothing in (8.19) and (8.20) which requires income to rise or fall monotonically as we move from one occupation to the next, so that we need no prior commitment as to which occupation is the higher-income one. The labeling is completely innocent.

By now, it should also be clear that we might mix conventional and dummy variables together in a single relationship. In particular, the use of dummy variables to represent phenomena which are not directly observable lends considerable flexibility to linear regression analysis in a wide variety of empirical problems. A judicious use of them can go far toward avoiding the rigidity of strict functional forms. A few examples should serve.

Suppose that we accept a linear formulation for the consumption–income relationship but believe that this function shifted during the war years (due to rationing, for example). Suppose that we know or are willing to assume that the shift was a parallel one, so that

$$E(y \mid X_1, X_2) = \begin{cases} \alpha_0 + \beta_1 X_1 & \text{if } X_2 \text{ is "peace,"} \\ \alpha_1 + \beta_1 X_1 & \text{if } X_2 \text{ is "war,"} \end{cases} \qquad (8.21)$$

where y = consumption, X_1 = income, and X_2 indicates whether peace or war prevailed. The linear formulation of this is

$$E(y \mid X_1, X_2) = \beta_0 + \beta_1 x_1 + \beta_2 x_2, \qquad (8.22)$$

where the definition of regressors is $x_1 = X_1$, $x_2 = 0$ in peace, 1 in war.

Suppose that we know, or are willing to assume, that the (average) relationship between a firm's investment and its profits is linear but also depends upon "management personality" (a permanent feature of the firm). On the assumption that the "personality" differences affect only the level of investment, the marginal propensity to invest out of profits being the same for all firms, we have the following form of the PRF:

$$E(y \mid X_1, X_2) = \beta_{0i} + \beta_1 X_1 \text{ for firm } i \qquad (i = 1, \ldots, I), \quad (8.23)$$

where $y =$ investment, $X_1 =$ profits, and X_2 simply labels the firms $i = 1, \ldots, I$. This translates into the single linear relationship

$$E(y \mid X_1, X_2) = \beta_{01} x_{01} + \cdots + \beta_{0I} x_{0I} + \beta_1 x_1, \qquad (8.24)$$

where the regressors are defined by $x_1 = X_1$, $x_{0i} = 1$ for firm i, 0 for other firms ($i = 1, \ldots, I$). Given a sample of IT observations on investment and profits for each of I firms in each of T years, we may apply linear regression to (8.24); see Greenberg (1964).

Suppose that we believe that in a consumption function the effect of the conditioning variable $X =$ income change is not symmetric, the effect of a unit change in X being dependent upon whether X is positive or negative. Such a feature plays a role in some dynamic theories of consumer demand. Then rather than simply entering $\beta_1 X$ into the PRF, we should define the two regressors

$$x_1 = \begin{cases} X & \text{if } X > 0, \\ 0 & \text{if } X \le 0, \end{cases} \qquad x_2 = \begin{cases} 0 & \text{if } X > 0, \\ X & \text{if } X \le 0, \end{cases}$$

and enter $\beta_1^* x_1 + \beta_2^* x_2$. This treatment may be interpreted as first defining the dummy variable $z = 1$ if $X > 0$, 0 if $X \le 0$, and then defining $x_1 = zX$ and $x_2 = (1 - z)X$.

8.4. TRANSFORMATIONS OF THE DEPENDENT VARIABLE

Thus far we have illustrated how the possibility of transforming the conditioning variables gives considerable flexibility to linear regression. Further flexibility is available if we consider transformations of the dependent variable as well.

The nonlinear function best known in economics is perhaps the *multilog* (= "Cobb-Douglas") function

$$Y = \beta_0 X_1^{\beta_1} \cdots X_K^{\beta_K}. \qquad (8.25)$$

This form has constant elasticities with respect to each of the conditioning variables: $(\partial Y / \partial X_k)(X_k / Y) = \beta_k$ ($k = 1, \ldots, K$). It has been very widely used in theoretical and empirical analyses of production, with Y being output and $X_1, \ldots, X_K$ being factor inputs. This function may be written as a linear (and additive) one by transforming the dependent variable along with the conditioning variables. Specifically, (8.25) may be written

$$y = (\log \beta_0) + \beta_1 x_1 + \cdots + \beta_K x_K \qquad (8.26)$$

where $y = \log Y$, $x_1 = \log X_1, \ldots,$ and $x_K = \log X_K$.

It would then appear that linear regression of the regressand y on the regressors $x_1, \ldots, x_K$ will be directly applicable in an empirical context. However, since the dependent variable has been transformed, some caution is required in specifying the stochastic model.

To investigate the issues, which are often overlooked, let us consider the case $K = 1$, where (8.25) reduces to

$$Y = \beta_0 X^{\beta_1}; \qquad (8.27)$$

the reader is invited to carry through the argument for the general case. We now consider a stochastic setting for (8.27). If we assume that $\beta_0 X^{\beta_1}$ is the PRF of Y, then $(\log \beta_0) + \beta_1 x$ will not be the PRF of y; that is, $E(Y \mid X) = \beta_0 X^{\beta_1}$ does not imply $E(y \mid X) = (\log \beta_0) + \beta_1 x$. After all, for nonlinear functions the expectation of the function is not the function of the expectation. Indeed, it can be shown that $E(y \mid X) < (\log \beta_0) + \beta_1 x$. For the expectation of the logarithm of a random variable must be less than the logarithm of the expectation of the random variable (unless the random variable has zero variance, in which case "less than" becomes "equal to"). This is an immediate consequence of the concavity of the logarithmic function, and may be demonstrated as follows. Let Y be a random variable which is always positive, with expectation $E(Y)$, and let $y = \log Y$. Taking the line tangent to $y = \log Y$ at the point $Y = E(Y)$ defines a new random variable,

$$z = \log E(Y) + [1/E(Y)][Y - E(Y)]. \qquad (8.28)$$

Since z is a linear function of Y, its expectation is readily seen to be

$$E(z) = \log E(Y) + [1/E(Y)]E[Y - E(Y)] = \log E(Y). \qquad (8.29)$$

Now, in view of the concavity of the logarithmic function, a line which is tangent to it lies above it everywhere (except at the point of tangency, of course). Therefore, the random variable y is always less than z, except when $Y = E(Y)$. If $y \leq z$ always, then certainly $y \leq z$ on the average, that is, $E(y) \leq E(z) = \log E(Y)$. The possibility $E(y) = E(z)$ being ruled out as soon as Y has a nonzero variance, we conclude that $E(\log Y) = E(y) <$ $E(z) = \log E(Y)$, as was to be demonstrated. While our demonstration has been done in terms of unconditional expectations, it clearly applies in terms of conditional expectations as well.

Thus the specification $E(Y \mid X) = \beta_0 X^{\beta_1}$ does not produce $E(y \mid X) = (\log \beta_0) + \beta_1 x$. Just what it does produce depends on other aspects of the model. Suppose we specify that

$$Y = \beta_0 X^{\beta_1} + u, \qquad (8.30)$$

where u is an additive disturbance distributed independently of X with expectation zero. Then we have $E(Y \mid X) = \beta_0 X^{\beta_1} + E(u \mid X) = \beta_0 X^{\beta_1}$, so that $\beta_0 X^{\beta_1}$ is the PRF of Y. But taking logarithms in (8.30) yields $y = \log Y = \log(\beta_0 X^{\beta_1} + u)$, which is a rather intractable expression that clearly does not give y a linear PRF; note that $\log(\beta_0 X^{\beta_1} + u) \neq (\log \beta_0) + \beta_1 x + (\log u)$.

Alternatively, suppose we specify that

$$Y = \beta_0 X^{\beta_1} \varepsilon, \qquad (8.31)$$

where ε is a multiplicative disturbance distributed independently of X with expectation *unity*. Then once again we have $\beta_0 X^{\beta_1}$ for the PRF of Y:

$$E(Y \mid X) = \beta_0 X^{\beta_1} E(\varepsilon \mid X) = \beta_0 X^{\beta_1} E(\varepsilon) = \beta_0 X^{\beta_1}, \qquad (8.32)$$

using the rule of (2.1), the independence of ε and X, and $E(\varepsilon) = 1$. Proceeding to take logarithms in (8.31), we obtain

$$y = (\log \beta_0) + \beta_1 x + (\log \varepsilon), \qquad (8.33)$$

where $\log \varepsilon$, like ε itself, is independent of X (and hence of x). To be sure, the disturbance in (8.33), $\log \varepsilon$, does not have a zero expectation—the concavity argument introduced above actually tells us that $E(\log \varepsilon) < \log E(\varepsilon) = \log 1 = 0$. Thus (8.33) is not suitable for linear regression. But now defining $\varepsilon^* = (\log \varepsilon) - E(\log \varepsilon)$ and also $\alpha = (\log \beta_0) + E(\log \varepsilon)$, we may rewrite (8.33) in a form which is suitable for linear regression:

$$y = \alpha + \beta_1 x + \varepsilon^*. \qquad (8.34)$$

In (8.34) the disturbance ε^* certainly has zero expectation—since $E(\varepsilon^*) = E[(\log \varepsilon) - E(\log \varepsilon)] = 0$—and is independent of x—since $\log \varepsilon$ is independent of x. We now have a linear PRF for y, namely,

$$E(y \mid x) = \alpha + \beta_1 x. \qquad (8.35)$$

Thus the multiplicative disturbance specification has some advantages over the additive disturbance specification for the multilog functions we are considering. It is, indeed, the prevalent one in practice; see, for example, Klein (1962, pp. 90–102). Before proceeding, it is interesting to note that under the multiplicative disturbance specification, not only the conditional expectation of Y, but also its conditional variance, varies with X. To see this, first subtract (8.32) from (8.31) to find $Y - E(Y \mid X) = \beta_0 X^{\beta_1}(\varepsilon - 1) = E(Y \mid X)(\varepsilon - 1)$. Then square and take the expectation conditional on X:

$$E\{[Y - E(Y \mid X)]^2 \mid X\} = E^2(Y \mid X)E[(\varepsilon - 1)^2 \mid X] = E^2(Y \mid X)\sigma^2, \quad (8.36)$$

where σ^2, the variance of ε, is the same for all X, since ε has been assumed to be independent of X. According to (8.36), the conditional variance of Y will vary with X, being proportional to the square of the conditional expectation of Y. This "heteroskedasticity" across the conditional distributions of Y may constitute an incidental virtue of the multiplicatve disturbance specification. For it seems that in empirical situations in which the multilog form is appropriate for the PRF, one frequently finds a systematic variation of the conditional variance as well. For example, upper-income groups show a wider variation in their expenditures on a commodity than do lower-income groups; see Prais and Houthakker (1955, pp. 55–56). Note that the conditional variance of $y = \log Y$ *is* assumed to be the same for all values of X; from (8.34) and (8.35) in the present specification, we have

$$E\{[y - E(y \mid x)]^2 \mid x\} = E(\varepsilon^{*2} \mid x) = E(\varepsilon^{*2}).$$

To pick up the main thread of the argument: We are proposing to attack the nonlinear PRF $E(Y \mid X)$ in (8.32) by means of the linear PRF $E(y \mid x)$ in (8.35). Given a sample Y_t, X_t $(t = 1, \ldots, T)$ we can construct the sample y_t, x_t $(t = 1, \ldots, T)$, and regress y on x, fitting $\hat{y}_t = a + b_1 x_t$ by least squares. Clearly a and b_1 are estimates of α and β_1, respectively. These least-squares estimates will have the usual desirable properties, on the understanding that the ε's (and hence the ε^*'s) are mutually independent, so that the stochastic specification of Section 7.2 applies to (8.34).

Does this estimation of $E(y \mid x)$ provide us an estimate of $E(Y \mid X)$, which after all was our original concern? The least-squares slope b_1 does estimate the elasticity in $E(Y \mid X)$, since that elasticity, β_1, is identical with the slope in $E(y \mid x)$. However, the least-squares intercept a estimates $\alpha = (\log \beta_0) + E(\log \varepsilon)$, while the constant in $E(Y \mid X)$ is β_0. In transforming the function we have transformed this parameter and therefore lack a direct estimate of it. In most economic applications it is the elasticity (elasticities in the case $K > 1$) which is important, so that the problem of the transformed intercept is not a serious one.

While our analysis of the multilog functional form has emphasized some problems, this should not distract us from the main point. Allowing for transformations of the dependent variable permits us to transform some nonlinear PRF's into linear ones, thus widening the class of models which can be attacked by linear regression. For further examples and references, see Draper and Smith (1966, pp. 133–134).

The burden of this chapter may now be restated. Suppose that we are interested in $E(Y \mid X_1, \ldots, X_J) = g(X_1, \ldots, X_J)$. Suppose further that there are known functions $y = h(Y)$ and $x_k = h_k(X_1, \ldots, X_J)$ $(k = 1, \ldots, K)$ such that $E(y \mid X_1, \ldots, X_J) = \beta_0 + \sum_{k=1}^{K} \beta_k x_k$. Then the sample $Y_t, X_{t1}, \ldots,$ X_{tJ} $(t = 1, \ldots, T)$ can be viewed as a sample $y_t, x_{t1}, \ldots, x_{t_K}$ $(t = 1, \ldots, T)$ and linear regression applied to estimate the transformed function. The parameters of the transformed PRF will be intimately related to those of the original PRF, so that, generally speaking, the estimates for the transformed PRF can be translated into estimates for the original PRF.

8.5. SUPPLEMENTARY REMARKS

1. An appreciation of the flexibility of linear regression analysis is best obtained by studying how it has been employed by economic researchers in a variety of empirical contexts. Surveying the current journal literature is recommended for this purpose. Reference may also be made to Ezekiel and Fox (1959, Chaps. 6, 14, 21) and Johnston (1963, pp. 44–52, 221–228).

2. While our nonlinear liquidity preference function (8.5) cannot be transformed into a linear regression, it may still be estimated by least squares. The normal equations, obtained by minimizing the residual sum of squares with respect to β_0, β_1, and β_2, will be nonlinear. This not only means that the computations are complicated but also that the theoretical analysis of Chapter 7 is not strictly applicable. That analysis relied heavily on the fact that the deviations of least-squares coefficients from the population parameters were linear functions of the disturbances. It is not hard to see that this feature gets lost when the normal equations are nonlinear. An instructive introduction to nonlinear regression may be found in Draper and Smith (1966, Chap. 10). A general conclusion which emerges from the analysis of nonlinear models is that least-squares estimation retains some, although not all, of its desirable statistical inference properties.

3. The use of dummy-variable regressors is closely related to the classical testing procedures of analysis of variance and covariance. This relation is discussed briefly by Goldberger (1964, pp. 227–231), in considerable detail by Watts (1964), and in rich empirical context by Kuh (1963, Chap. 5). One very fruitful use of dummy variables in conjunction with ordinary variables is concerned with seasonal variations; see Klein (1962, pp. 35–39). Indeed, the regression method for seasonal adjustment of economic time series now appears to have considerable advantages over the traditional moving-average technique (and its complicated modern variants); see Jorgenson (1964).

4. The problem concerning estimation of the constant β_0 in the multilog function (8.31) can be resolved in a simple manner if the disturbance ε in (8.31) is lognormally distributed. In that case, it can be shown that the "nuisance parameter" $E(\log \varepsilon)$ equals minus one half the variance of $\log \varepsilon$. Also, $\log \varepsilon$ and ε^* differ by a constant and hence have the same variance. The variance of ε^* is estimable in the usual way from the error sum of squares in the logarithmic regression (8.34); thus $E(\log \varepsilon)$ is estimable. This estimate of $E(\log \varepsilon)$ can be combined with a, the estimate of α, to obtain an estimate of β_0. Procedures for doing this, and further aspects of the model, are developed in Goldberger (1968).

Chapter 9

Choice of Functional Form

9.1. INTRODUCTION

$\mathbf{A}$ WIDE VARIETY OF FUNCTIONAL FORMS FOR THE PRF can be handled within the framework of linear regression analysis. It remains to consider the bases for choosing among the many available alternatives in particular economic contexts. What we have seen in essence is that if we know or are willing to assume a specific functional form (along with some stochastic specification of the disturbance distribution as in Chapter 7), there exist rather well-defined statistical procedures for estimating the unknown parameters of that function. But what is the basis for our "knowledge" or "willingness to assume" a particular functional form for $E(Y \mid X_1, \ldots, X_J)$? Indeed, what is the basis for our "knowledge" or "willingness to assume" that the relevant conditioning variables are $X_1, \ldots, X_J$—just those and no others?

Unfortunately, no clear-cut answer to these questions exists. There is no magic rule which will tell us which is the appropriate model to employ for a given empirical problem. Nevertheless, it is possible to formulate some guidelines, or partial criteria, which the empirical researcher is well advised to keep in mind. While these cannot resolve the deep problem of choice, they may go some ways toward narrowing down the range of choice in a rational way.

At the outset we may dispose of two extreme positions—the "theorist" and the "empiricist." At one extreme it may be argued that economic theory should be relied on exclusively, to determine which variables are relevant to the PRF and in what form they should enter the PRF. At the other extreme it may be argued that we should let the data speak for themselves, that is, choose the functional form which best fits the sample data. Neither of these positions is in fact tenable—nor is either seriously entertained by researchers. With respect to the theorist extreme, we note first that economic theory is simply not as generous as that position would require. At best economic theory may suggest which variables are definitely relevant, which others are possibly relevant, and perhaps some broad features of the functional form. But it is not very usual for economic theory to lead to a specific functional form. And, of course, the purpose of the empirical investigation may be precisely to decide between two or more conflicting economic theories. Furthermore, the variables considered in the theory may be unobserved (or observed only approximately) in our sample. In that event some allowance is required if we are to derive an operational functional form from the abstractions of theory. With respect to the empiricist extreme, we note first that data never really speak for themselves. The theory of statistical inference, while it may enable us to choose among a narrow set of alternative hypotheses, is simply not powerful enough to make a choice among an undefined open-ended variety of hypotheses. As to "fitting the data best," the mere fact that with sufficient ingenuity and persistence any of us could come up with a "perfect" explanation of any set of observations on Y should dissuade us from relying on that criterion exclusively. (In technical terms, any given $T \times 1$ vector can be expressed as an *exact* linear function of *any* set of T linearly independent $T \times 1$ vectors.) For none of us would expect that a function painstakingly selected to explain all the peculiarities of a given sample would have any transferability to another sample. Our objective, presumably, is not to explain "perfectly" the features of a particular historical phenomenon, but rather to come up with an explanation which has some carryover to related phenomena, past or future.

It should be clear then that neither prior knowledge nor the sample can by itself resolve the problem of choice, but that some mixture of the two is required. The productive researcher will want to draw upon both "theory" and "fact" in his investigations. One is tempted to say that it is just this necessity to blend theory and fact judiciously that makes empirical economics an art and not a science. But this is misleading, since the same necessity, in greater or lesser degree, prevails in all sciences.

Now, we are considering empirical investigations which have as their objective the ascertainment of permanent features of economic behavior (that is, the reliable estimation of population parameters) rather than the mere fitting of a given body of data (that is, the computation of sample statistics). This suggests that the ultimate criterion for choice among alternative functional forms would be based upon predictive power, the best model being that which predicts the best. Indeed, there is general agreement on this—in some version—as the ultimate criterion. But even so, the researcher is faced with the problem of making a tentative choice among alternative models before all the predictive evidence is available.

9.2. ECONOMIC-THEORETICAL CRITERIA

Whenever possible, economic theory should be exploited in empirical research. As noted previously, economic theory often implies certain broad features of the PRF—for example, marginal cost functions which fall and then rise, Engel curves which have an initial income level and a saturation expenditure level, consumption functions which have declining marginal propensities to consume. It is clearly desirable to consider functional forms which allow those theoretical features to manifest themselves. In this connection, we should interpret economic theory broadly to include the results of prior empirical investigations, so that if some functional form has proved useful in earlier work on related data it should be considered again.

In addition to such broad features, economic theory often places restrictions upon the form of the function. A familiar example is the requirement in classical consumer theory that a demand function should be homogeneous of degree zero in money income and prices. Then we may consider only functions which have this property or, to put it another way, reduce the number of explanatory variables by expressing them relative to one numeraire price. Furthermore, economic theory often places some restrictions on the signs or orders of magnitude of coefficients.

For an instructive application of economic theory in choosing among alternative functional forms, see Prais and Houthakker (1955, Chap. 2, pp. 82–88).

It should be kept in mind that the relevance of economic theory to an empirical problem may be open to question. For example, is the classical equilibrium consumer demand theory relevant to a single cross section? A continuing stimulus to the development of economic theory has been precisely the appearance of empirical phenomena which seemed inconsistent with the previously existing theory. Malinvaud (1966, Chap. 4) provides an extraordinarily illuminating discussion of this.

9.3. FORMAL STATISTICAL TESTS OF HYPOTHESES

Formal statistical tests of hypotheses—such as those based on the Student t and Snedecor F distributions of certain sample statistics—are available for use in selecting among competing functions. Thus if we are uncertain whether or not to include a particular regressor, we may tentatively include it, and then test the hypothesis that its population regression coefficient is zero. The regressor may represent an additional conditioning variable—for example, liquid assets in addition to income in a consumption function—or may represent a more complicated functional form for the same conditioning variable—for example, the quadratic term in (8.3) and (8.4).

The choice between two alternative theories is facilitated if they can both be treated as special cases of a more general theory. We can then estimate the PRF corresponding to the general theory and test which of the special cases is appropriate. Malinvaud (1966, pp. 134–135) offers an example: Suppose that we are hesitating between

$$C_t = \beta_0 + \beta_1 Y_t + \beta_2 Y_{t-1} + \varepsilon_t \tag{9.1}$$

and

$$C_t = \beta_0^* + \beta_1^* Y_t + \beta_2^* C_{t-1} + \varepsilon_t^*, \tag{9.2}$$

where C = consumption and Y = income. Then we might formulate the equation

$$C_t = \alpha_0 + \alpha_1 Y_t + \alpha_2 Y_{t-1} + \alpha_3 C_{t-1} + \varepsilon_t^{**}, \tag{9.3}$$

which contains (9.1) as the special case $\alpha_3 = 0$ and (9.2) as the special case $\alpha_2 = 0$. Estimation of (9.3) should then enable us to test which of the two null hypotheses is accepted and which rejected, and thus to decide which of the two forms (9.1) or (9.2) is appropriate.

Some comments on the use of such formal tests are in order:

1. They are particularly suited for choosing among a small set of well-defined alternatives where those can be brought within the framework of a single general model. There are inherent difficulties if we attempt to test a large number of alternative hypotheses one by one.

2. If we are interested in testing whether a conditioning variable enters the PRF, and this conditioning variable is represented by several different regressors, then the appropriate test is a joint one, not a series of simple ones. For example, in the quadratic model (8.3) and (8.4) if we want to test

the hypothesis that $E(y \,|\, X)$ does not depend on X, the appropriate null hypothesis is $\beta_1 = 0 = \beta_2$ and not the two separate hypotheses $\beta_1 = 0$ and $\beta_2 = 0$.

3. Strictly speaking, these procedures do not test whether a conditioning variable is relevant for the PRF, but rather whether it is relevant in the form in which it has been entered. For example, it is logically possible to accept the hypothesis $\beta_1 = 0$ in $E(y \,|\, X) = \beta_0 + \beta_1 X$ and to reject the hypothesis $\alpha_1 = 0$ in $E(y \,|\, X) = \alpha_0 + \alpha_1 \log X$. As a practical matter, however, if a variable really has nonlinear effect, we are likely to find some effect even if we enter it only linearly. This is attributable to the fact that over short ranges most functions are reasonably well approximated by linear functions.

4. The power of these tests is inhibited by the prevalence of multicollinearity, as indeed is any procedure for discriminating among alternatives. For example, consider the consumption functions (9.1) to (9.3). Since Y_{t-1} and C_{t-1} are likely to have moved closely together in a (time-series or cross-section) sample, it is quite possible that both simple alternatives $\alpha_2 = 0$ and $\alpha_3 = 0$ will be accepted (the joint hypothesis $\alpha_2 = 0 = \alpha_3$ being rejected, say), and the choice will not have been resolved. In that event we may decide that both the lagged variables should be retained—that is, we may be content to revise our theory to allow for the possibility that both lagged consumption and lagged income influence current consumption. Indeed, it seems that the conflict between alternative theories is often a false one, in the sense that each emphasizes some, and neglects other, relevant elements of an adequate theory. If we insist on making a choice, we may choose between (9.1) and (9.2) on the basis of their respective R^2's, which amounts to rejecting that one of the two null hypotheses $\alpha_2 = 0$ or $\alpha_3 = 0$ which leads to the largest reduction of R^2 in (9.3).

9.4. GOODNESS OF FIT AND R^2

A very common procedure for choice among alternative functions is of course to select the one which fits the best—that is, the one which has the highest R^2. A heuristic justification for this procedure would rely on the intimate connection between R^2 and the F statistic which enters the formal tests mentioned in Chapter 8 and Section 9.3. In some contexts this justification can be made rigorous.

But while few of us would regret obtaining a high R^2, there are serious objections to relying exclusively or very heavily on this empiricist criterion. Some comments are in order:

1. Since our objective is to obtain reliable estimates of population parameters and not merely to account for the variation in a single sample, we should be wary of choosing functions which have a high R^2 but many "nonsignificant" coefficients, or coefficients whose signs or magnitudes have no theoretical support.

2. Since it is always possible to increase R^2 by adding additional regressors, we should be wary of functions which attain a high R^2 by the use of many regressors. An adjusted coefficient of determination may be more appropriate in choosing among functions which use differing numbers of regressors. The traditional adjustment is to take instead of R^2 the sample statistic

$$\bar{R}^2 = 1 - (1 - R^2)\frac{T-1}{T-K-1} = R^2 - \frac{K}{T-K-1}(1 - R^2), \quad (9.4)$$

which clearly penalizes functions having a higher K (particularly when T is small).

3. Since R^2 measures the proportion of the variation of the regressand which is accounted for by the SRF, the R^2's of functions which have different regressands are not directly comparable. Suppose we wish to choose between $E(Y|X) = \beta_0 + \beta_1 X$ and $E(y|X) = \beta_0^* + \beta_1^* X$, where $y = \log Y$. Suppose further that we agree to choose the one which does a better job of accounting for the sample variation of Y. Now while the R^2 of the first function measures the proportion of the variation of Y explained, the R^2 of the second function measures the proportion of the variation of $\log Y$ explained, which is not the same animal. To obtain a comparable measure from the second function we may proceed as follows. Compute the $\hat{y}_t$, the calculated values from the second function; take their antilogs, $\hat{Y}_t^* = \text{antilog } \hat{y}_t$. These are obviously estimates of the absolute rather than logarithmic values. Then compute the R^2 between Y_t and $\hat{Y}_t^*$—this is comparable to the R^2 of the first function, which is the R^2 between Y_t and $\hat{Y}_t$.

9.5. RESIDUAL PATTERNS

A study of the residuals from an SRF is often useful in suggesting additional regressors (either to represent new conditioning variables or to represent new functional forms for the conditioning variables already used). The general idea is to look for patterns in the residuals—that is, for relationships between them and other variables. This may be done by graphical inspection, by computing the mean residual for groups defined by values of the potential regressor or conditioning variable, or by a formal regression on the potential variable or variables. Draper and Smith (1966, Chap. 3) offer some useful hints.

Such a search, provided it is guided by economic-theoretical considerations, is to be encouraged. Some comments are in order:

1. The tools employed in the examination of residuals are in essence forms of "stepwise regression." As the discussion in Chapter 3 suggests —see especially (3.32) and (3.56)—"stepwise regression" is likely to understate the relevance of the potential regressor in the sense of "giving it less credit" than it would have received had it been included in the relationship in the first place. Nevertheless, the stepwise procedure may be defended on the grounds of research strategy: The researcher is likely to have oversimplified his model at first in fitting the original function. In any event, after making the search, he is free to fit the revised functional form.

2. The traditional formal hypothesis testing procedures are designed to test hypotheses proposed in advance of examination of the sample and do not carry over to hypotheses developed after examination of the sample. It follows that standard errors and similar statistics, computed in the conventional manner for an SRF which has been chosen after exploring several alternatives with the same body of data, can be taken only as indicative and have no rigorous justification. Intuitively it is clear that the conventional measures will overstate the precision which attaches to a set of estimates which have been chosen on the basis of goodness of fit from among a batch of alternatives. Unfortunately, few theoretical results are as yet available to indicate just what test procedures or imprecision measures are appropriate when the present multistage procedure is followed. The problem of "pretesting bias" remains troublesome; for some analysis of this bias see Bancroft (1944).

3. Residual patterns may well be more useful than R^2 in suggesting improvements in the functional form. Suppose we have obtained a high R^2 with the linear form $\hat{y} = b_0 + b_1 X$. If the residuals from this regression show a striking pattern, being low for low values of X, high for intermediate values of X, and low again for high values of X, then the addition of a quadratic term should definitely be considered; see Prais and Houthakker (1955, pp. 51–55).

4. Residual patterns are often suggestive of a nonhomogeneity in the sample in the sense that two or more groups with different behavior patterns have been lumped together. For example, if we fit a production function to a group of firms, obtain a high R^2, but find that the large firms

tend to have negative residuals while the small firms have positive ones, we should suspect some misspecification of the production function. A similar phenomenon in consumption functions was among the early pieces of evidence offered in support of the permanent income hypothesis; see Friedman (1957, pp. 48–51).

5. As thus far described, the search for residual patterns involves a questioning of the original specification of the form of the PRF. But such a search is also important when other stochastic specifications are open to question. For example, the stochastic model developed in Chapter 7 included the assumptions of "homoskedasticity"—$E(\varepsilon_t^2)$ constant for all t—and "nonautocorrelatedness"—$E(\varepsilon_t \varepsilon_s) = 0$ for $t \neq s$. If, as is often the case, there is reason to doubt these assumptions, they should be doubted. Procedures for examining their validity frequently rely upon the sample residuals, which are, after all, estimates of the population disturbances. Such testing procedures, and methods for parameter estimation in the presence of heteroskedasticity and/or autocorrelatedness, are considered in Johnston (1963, pp. 207–211, Chap. 7), Goldberger (1964, pp. 231–246), and Malinvaud (1966, pp. 254–258, Chap. 13).

9.6. MULTICOLLINEARITY

Multicollinearity inhibits our ability to discriminate among competing alternative functional forms. If two regressors move very closely together over the sample observations it will be very difficult, if not impossible, to decide on the basis of the sample which of them is the appropriate one for the PRF.

A few comments are in order:

1. Alternative functional forms may fit the sample data equally well and yet have quite different theoretical implications. Their predictive record may be sharply different if the close connection between their regressors does not persist in the future. Their policy implications may be quite different, particularly since policy proposals often involve destroying an existing relation. (*Example:* Gross income and disposable income may perform equally well as explainers of consumption over a period in which the tax rate was fixed, but a changed tax policy will precisely destroy the previous relation between them.)

2. The prevalence of severe multicollinearity in time-series samples has made cross-section samples more attractive to researchers. Cross sections are often richer in terms of independent variation of the explanatory

variables and, in any event, are larger in size. This case for microdata as opposed to macrodata has been put strongly by Orcutt (1962). The low R^2's typically found in cross-section studies ought not to reduce the attraction for those who are interested in obtaining reliable estimates of population parameters and not merely good fits. Macrodata involves grouping or averaging of microdata and it can be shown that such groupings or averaging has a built-in tendency to raise R^2's; see Cramer (1964).

3. When severe multicollinearity is present in the sample, the value of prior information—economic theory or previous empirical results—is enhanced. In Section 6.3 we had some examples of this. Also the use of rather rigid functional forms may be required. It is sometimes argued that since any function can be approximated by a (sufficiently high-order) polynomial in view of Taylor's theorem, the practice of including several powers of a regressor should be encouraged. However, there are no clear-cut means of deciding what order polynomial to take. Also, and more important, there is likely to be severe multicollinearity among a set of regressors such as $X, X^2, X^3, \ldots$. Instead, one should consider some simple linear or curvilinear form which uses less regressors.

9.7. SIMPLICITY

Implicit in much of the previous discussion of this chapter is the idea that simplicity is a virtue. Other things being equal, a simple functional form is to be preferred to a complex one. This dictum is in accord with a general principle of philosophy and scientific method known as *Occam's razor*: Do not compound hypotheses unnecessarily. The importance of this principle is well known in economic theory, where credit goes to those theories which can account for many facts reasonably well with a minimum of assumptions. This emphasis on simplicity is the reason linear functions are so extensively used in economics.

The development of distributed lag models provides an interesting historical example of the role played by the simplicity criterion in empirical economic research. Early studies would relate the dependent variable in period t to the explanatory variable in the same period,

$$y_t = \beta x_t + \varepsilon_t, \tag{9.5}$$

where here and in what follows the intercept is omitted for convenience. This implied that the response to change in x was fully made in the same period in which the change took place. Since this did justice to neither the

theoretical nor empirical evidence on delayed reactions, researchers turned
to functions which included a fixed one-period lag, say,

$$y_t = \beta x_{t-1} + \varepsilon_t \tag{9.6}$$

or sometimes

$$y_t = \beta x_t + \beta_1 x_{t-1} + \varepsilon_t. \tag{9.7}$$

But these did not seem adequate either, since real-world responses appear
to be stretched out over more than two periods. This led to the concept
of distributed lag functions which distribute the response over many
periods, say,

$$y_t = \beta_0 x_t + \beta_1 x_{t-1} + \cdots + \beta_K x_{t-K} + \varepsilon_t. \tag{9.8}$$

The difficulties with this approach were soon realized: The successive
lagged values of x tend to be highly multicollinear, limited sample sizes
prevent the use of many regressors, and in any event once having admitted
the possibility of longer delays, it becomes hard to find a reason for cutting
off the sequence at any particular value of K. It was then recognized that
these problems could be avoided and a major simplification accomplished
if one was willing to impose some restriction on the values of the β's,
that is, to make some assumptions about the relative strength of the
influence of the lagged values of x upon y_t. Presumably there is general
agreement that the influence declines as we go further back in time, at
least after some point. A natural way of capturing this idea is the Koyck-
Nerlove geometrically declining weights assumption,

$$\beta_k = \lambda^k \beta \qquad (0 < \lambda < 1). \tag{9.9}$$

That this achieves a powerful simplification is readily seen. Suppose we
allow for an indefinite series of delayed responses,

$$y_t = \beta x_t + \beta_1 x_{t-1} + \beta_2 x_{t-2} + \cdots + \varepsilon_t \tag{9.10}$$

but impose (9.9):

$$y_t = \beta x_t + \lambda \beta x_{t-1} + \lambda^2 \beta x_{t-2} + \cdots + \varepsilon_t. \tag{9.11}$$

If we lag (9.11) by one period we have

$$y_{t-1} = \beta x_{t-1} + \lambda \beta x_{t-2} + \cdots + \varepsilon_{t-1}; \tag{9.12}$$

multiplying (9.12) through by λ, subtracting from (9.11), and rearranging gives

$$y_t = \beta x_t + \lambda y_{t-1} + \varepsilon_t^*, \tag{9.13}$$

where $\varepsilon_t^* = \varepsilon_t - \lambda \varepsilon_{t-1}$. Clearly (9.13) is a simple form—it involves only two regressors—and yet allows for an indefinite series of delayed reactions. Functions of this form have proved very useful in empirical research; for illustrations, a discussion of problems arising in estimation of (9.13), and alternative patterns of distributed lags, see Malinvaud (1966, pp. 111–117, Chap. 15).

9.8. SUPPLEMENTARY REMARKS

1. As was promised, the scattered guidelines mentioned in this chapter cannot be collected into a single rule for optimal selection of functional forms. Indeed, some of the guidelines will often be in conflict. It may be said that the art of empirical research lies in knowing how to pick and weigh such guidelines. The difference between a mechanical and an artistic use of regression analysis is considerable in terms of the contribution to economic science.

2. It will have been recognized that economic theory and sample information are substitutes. When we have a rich body of economic theory to guide us, then even a limited sample can provide reliable estimates of characteristics of the stochastic relationship in which we are interested. When we have a rich sample—large in size and independent variation of the explanatory variables—we can experiment and choose among a large variety of theories.

3. An appreciation of the criteria for choosing among alternative functional forms is best obtained by studying how they are employed by economic researchers in a variety of empirical contexts. A classic source is Prais and Houthakker (1955, Chap. 7).

Bibliography

Allen, R. G. D., *Mathematical Economics*, New York: Macmillan, 1956.

Bancroft, T. A., "On Biases in Estimation Due to the Use of Preliminary Tests of Significance," *Annals of Mathematical Statistics*, Vol. 15, June 1944, pp. 190–204.

Courant, R., *Differential and Integral Calculus*, Vol. II, New York: Wiley-Interscience, 1936.

Cramér, H., *Mathematical Methods of Statistics*, Princeton, N.J.: Princeton University Press, 1951.

Cramer, J. S., "Efficient Grouping, Regression and Correlation in Engel Curve Analysis," *Journal of the American Statistical Association*, Vol. 59, March 1964, pp. 233–250.

Draper, N. R., and H. Smith, *Applied Regression Analysis*, New York: Wiley, 1966.

Ezekiel, M., and K. A. Fox, *Methods of Correlation and Regression Analysis*, Third Edition, New York: Wiley, 1959.

Friedman, M., *A Theory of the Consumption Function*, Princeton, N.J.: Princeton University Press, 1957.

Gay, H. M., *Analytic Geometry and Calculus*, New York: McGraw-Hill, 1950.

Goldberger, A. S., *Econometric Theory*, New York: Wiley, 1964.

Goldberger, A. S., "On the Interpretation and Estimation of Cobb-Douglas Functions," *Econometrica*, Vol. 36, July–October, 1968.

Goldberger, A. S., and D. B. Jochems, "Note on Stepwise Least Squares," *Journal of the American Statistical Association*, Vol. 56, March 1961, pp. 105–110.

Greenberg, E., "A Stock-Adjustment Investment Model," *Econometrica*, Vol. 32, July 1964, pp. 339–357.

Griliches, Z., "Specification Bias in Estimates of Production Functions," *Journal of Farm Economics*, Vol. 39, February 1957, pp. 8–20.

Johnston, J., *Statistical Cost Analysis*, New York: McGraw-Hill, 1960.

Johnston, J., *Econometric Methods*, New York: McGraw-Hill, 1963.

Jorgenson, D., "Minimum Variance, Linear, Unbiased Seasonal Adjustment of Economic Time Series," *Journal of the American Statistical Association*, Vol. 59, September 1964, pp. 681–724.

Klein, L. R., "Estimating Patterns of Savings Behavior from Sample Survey Data," *Econometrica*, Vol. 19, October 1951, pp. 438–454.

Klein, L. R., *An Introduction to Econometrics*, Englewood Cliffs, N.J.: Prentice-Hall, 1962.

Klein, L. R., and A. S. Goldberger, *An Econometric Model of the United States 1929–1952*, Amsterdam: North-Holland, 1955.

Koutsoyianni-Kokkova, A., *An Econometric Study of the Leaf Tobacco Market of Greece*, Athens: 1962.

Kuh, E., *Capital Stock Growth: A Micro-Econometric Approach*, Amsterdam: North-Holland, 1963.

Malinvaud, E., *Statistical Methods of Econometrics*, Chicago: Rand McNally, 1966.

Mood, A. M., and F. A. Graybill, *Introduction to the Theory of Statistics*, Second Edition, New York: McGraw-Hill, 1963.

Orcutt, Guy H., "Microanalytic Models of the United States Economy: Need and Development," *American Economic Review*, Vol. 52, May 1962, pp. 229–240.

Prais, S. J., and H. S. Houthakker, *The Analysis of Family Budgets*, New York: Cambridge University Press, 1955.

Scheffé, H., *The Analysis of Variance*, New York: Wiley, 1959.

Theil, H., "Specification Errors and the Estimation of Economic Relationships," *Review of the International Institute of Statistics*, Vol. 25, 1957, pp. 41–51.

Watts, H. W., "An Introduction to the Theory of Binary Variables (or All About Dummies)," Systems Formulation and Methodology Workshop Paper 6404, Social Systems Research Institute, University of Wisconsin, Madison, 1964.

Watts, H. W., "The Test-O-Gram: A Pedagogical and Presentational Device," *American Statistician*, Vol. 19, October 1965, pp. 25–28.

Wold, H. O. A., "Forecasting by the Chain Principle," Chap. 1, pp. 5–36, in *Econometric Model Building*, H. O. A. Wold, editor, Amsterdam: North-Holland, 1964.

Index